Advanced Materials and Nanotechnology for Sustainable Energy and Environmental Applications

Advanced Materials and Nanotechnology for Sustainable Energy and Environmental Applications

Editors

Angela Malara
Patrizia Frontera

MDPI • Basel • Beijing • Wuhan • Barcelona • Belgrade • Manchester • Tokyo • Cluj • Tianjin

Editors
Angela Malara
Mediterranea University of
Reggio Calabria
Italy

Patrizia Frontera
Mediterranea University of
Reggio Calabria
Italy

Editorial Office
MDPI
St. Alban-Anlage 66
4052 Basel, Switzerland

This is a reprint of articles from the Special Issue published online in the open access journal *Applied Sciences* (ISSN 2076-3417) (available at: https://www.mdpi.com/journal/applsci/special_issues/Materials_Nanotechnology_Energy_Environmental).

For citation purposes, cite each article independently as indicated on the article page online and as indicated below:

LastName, A.A.; LastName, B.B.; LastName, C.C. Article Title. *Journal Name* **Year**, *Volume Number*, Page Range.

ISBN 978-3-0365-5229-3 (Hbk)
ISBN 978-3-0365-5230-9 (PDF)

© 2022 by the authors. Articles in this book are Open Access and distributed under the Creative Commons Attribution (CC BY) license, which allows users to download, copy and build upon published articles, as long as the author and publisher are properly credited, which ensures maximum dissemination and a wider impact of our publications.

The book as a whole is distributed by MDPI under the terms and conditions of the Creative Commons license CC BY-NC-ND.

Contents

About the Editors vii

Angela Malara and Patrizia Frontera
Special Issue on Advanced Materials and Nanotechnology for Sustainable Energy and Environmental Applications
Reprinted from: *Appl. Sci.* **2022**, *12*, 7440, doi:10.3390/app12157440 1

Filippo Fazzino, Emilia Paone, Altea Pedullà, Francesco Mauriello and Paolo S. Calabrò
A New Biorefinery Approach for the Full Valorisation of Anchovy Residues: Use of the Sludge Generated during the Extraction of Fish Oil as a Nitrogen Supplement in Anaerobic Digestion
Reprinted from: *Appl. Sci.* **2021**, *11*, 10163, doi:10.3390/app112110163 5

**Antonella Satira, Emilia Paone, Viviana Bressi, Daniela Iannazzo, Federica Marra,
Paolo Salvatore Calabrò, Francesco Mauriello and Claudia Espro**
Hydrothermal Carbonization as Sustainable Process for the Complete Upgrading of Orange Peel Waste into Value-Added Chemicals and Bio-Carbon Materials
Reprinted from: *Appl. Sci.* **2021**, *11*, 10983, doi:10.3390/app112210983 15

Andrea Gnisci, Antonio Fotia, Lucio Bonaccorsi and Andrea Donato
Effect of Working Atmospheres on the Detection of Diacetyl by Resistive SnO_2 Sensor
Reprinted from: *Appl. Sci.* **2022**, *12*, 367, doi:10.3390/app12010367 31

**Emanuela Mastronardo, Elpida Piperopoulos, Davide Palamara, Andrea Frazzica and
Luigi Calabrese**
Morphological Observation of LiCl Deliquescence in PDMS-Based Composite Foams
Reprinted from: *Appl. Sci.* **2022**, *12*, 1510, doi:10.3390/app12031510 41

**Francesca Nardelli, Emilia Bramanti, Alessandro Lavacchi, Silvia Pizzanelli,
Beatrice Campanella, Claudia Forte, Enrico Berretti and Angelo Freni**
Thermal Stability of Ionic Liquids: Effect of Metals
Reprinted from: *Appl. Sci.* **2022**, *12*, 1652, doi:10.3390/app12031652 55

Lucio Bonaccorsi, Andrea Donato, Antonio Fotia, Patrizia Frontera and Andrea Gnisci
Competitive Detection of Volatile Compounds from Food Degradation by a Zinc Oxide Sensor
Reprinted from: *Appl. Sci.* **2022**, *12*, 2261, doi:10.3390/app12042261 67

Ali Elrashidi
Light Harvesting in Silicon Nanowires Solar Cells by Using Graphene Layer and Plasmonic Nanoparticles
Reprinted from: *Appl. Sci.* **2022**, *12*, 2519, doi:10.3390/app12052519 77

About the Editors

Angela Malara

Angela Malara is a Researcher at the Department of Civil Engineering, Energy, Environment and Materials of the Mediterranean University of Reggio Calabria (Italy). Her research activity is focused on the development, the synthesis and the characterization of innovative nanostructured materials for energy, catalytic, environmental and sensing applications. She is the author of publications in international journals, national and international conference proceedings and co-inventor of one European patent.

Patrizia Frontera

Patrizia Frontera is an Associate Professor at the Department of Civil Engineering, Energy, Environment and Materials of the Mediterranean University of Reggio Calabria (Italy). She obtained her Ph.D. in Chemical and Materials Engineering from the University of Calabria. Her research interests are related to the evaluation of catalytic activity of micro-mesoporous materials, hydrogen production from hydrocarbons, exploitation of CO_2 for output of value-added products and investigations into nanostructures fabricated by electrospinning for employment in fuel cells, sensing and electrocatalysis. She is the author of more than seventy papers published in international journals, one European patent and about one hundred national and international conference proceedings.

Editorial

Special Issue on Advanced Materials and Nanotechnology for Sustainable Energy and Environmental Applications

Angela Malara * and Patrizia Frontera *

Department of Civil, Energy, Environment and Materials Engineering, Mediterranea University of Reggio Calabria, 89124 Reggio Calabria, Italy
* Correspondence: angela.malara@unirc.it (A.M.); patrizia.frontera@unirc.it (P.F.)

1. Introduction

Materials play a very important role in the technological development of a society, greatly impacting people's daily lives. Indeed, the development of innovative materials and the enhancement of existing ones are closely linked to the continuous demand for more advanced and sophisticated applications. Although aspects related to the study, the synthesis and the applications of materials are of interdisciplinary interest, in the last few years, great attention has been paid to the development of advanced materials for environmental preservation and sustainable energy technologies.

This Special Issue aimed to cover the current design, synthesis, and characterization of innovative advanced materials and nanotechnologies, as well as novel applications able to offer promising solutions to the these pressing themes.

2. Sustainable Energy and Environmental Applications

The development of new materials and processes, able to complete the approaching energetic transition challenge, aiming to produce, convert, harvest, transport, and store clean energy [1–3], is strictly related to global environmental issues.

Indeed, the limited reserves of fossil fuels, together with their high environmental impact on greenhouse gas emissions and the growth of the global energy demand, have raised the ambitious goal to produce efficient materials for energy-related fields, able at the same time to respond to sustainable and environmentally friendly principles [4,5]. The possibility to undertake a virtuous circuit that is able to give new life to waste products is also a possibility that well fits these objectives [6,7]. In order to reach these targets, multi-disciplinary expertise is required and encouraged.

In this view, in this Special Issue, novel, advanced and improved materials were fully investigated and applied to improve energy systems, sensing devices and waste valorization for fuel production.

Nardelli et al. [8] employed a promising ionic liquid as an advanced heat transfer fluid in solar thermal energy applications. Thanks to their high heat capacity, low melting point and relatively high density in the typical operating conditions of solar thermal energy systems, ionic liquid can be used in this field, but needs to be opportunely selected. Authors investigated the thermal stability and corrosion effects of these compounds, developing a promising method for their selection with a suitable lifetime able to meet the durability requirements of commercial and industrial solar thermal applications [8].

A composite material based on silicone vapor-permeable foam filled with lithium chloride salt was investigated for low-temperature heat storage applications by Mastronardo et al. [9]. Thermal energy storage is regarded an efficient and effective means to store solar energy when it is available in excess. In this process, heat is transferred to the material that, due to the dehydration reaction, is stored as long as the salt is in the dehydrated form. When heat is required, water is made accessible to the salt, reversible hydration takes place and heat is released, making the heat discharge available and controllable on demand. Lithium chloride salt was used for this purpose, but, being known for its deliquescent effect, it was incorporated in a composite matrix, characterized, and tested under hydration/dehydration cycles. The results showed an effective embedding of the material, which limited the salt release when overhydrated, thus making it a competitive candidate in the related application field [9].

A light harvesting system was presented by Elrashidi [10]. Silicon nanowires were coated with a graphene layer and plasmonic nanoparticles were distributed on the top surface of the silicon substrate layer to realize a solar cell. The proposed structure was used for efficient light harvesting in the visible and near-infrared regions. The performance of the solar cell was tested for different nanoparticle materials and dimensions, as well as for different solar cell structures, reporting improved results compared to similar reported systems [10].

Bonaccorsi et al. [11] presented a metal oxide sensor based on zinc oxide for possible use in the monitoring of low concentrations of volatile organic compounds, that are generally released during the phenomenon of food degradation. In particular, the device was able to detect the target volatile, hexanal, with minimum interference from all the others. In this way the material was proved to function as a selective gas sensor, giving an important indication of the quality and conservation of meat [11].

Similarly, Gnisci et al. [12] used a nanostructured material based on tin oxide to develop low cost and real-time resistive sensors useful in the monitoring of the fermentation process and storage of many foods and beverages. In particular, the effect of different working atmospheres was taking into account for the detection of gaseous diacetyl, generally produced during these processes and undesired over a specific threshold. The sensor showed a low detection limit, good selectivity and low response/recovery times [12].

The approach of waste valorisation was pursued instead by Fazzino et al. [13], whose study aimed to address the full valorization of anchovies in order to extract fish oil, an omega-3 source, and to produce biomethane through anaerobic digestion. In addition, the proposed extraction was a green-solvent process [13].

Finally, Satira et al. [14], starting from the hydrothermal carbonization of orange peel waste, presented a simple and green protocol to obtain hydrochar and high-added value products, such as 5-hydroxymethylfurfural, furfural, levulinic acid and alkyl levulinates. Numerous process variables were investigated in order to find the optimum conditions that maximized the yields of products resulting from waste valorisation [14].

3. Concluding Remarks

As proved by the authors, there are many facets of the same issue and many very different ways to contribute to improve clean energy technologies and environmental sustainability. The two aspects are strictly related and, as also shown by the latest trends in the scientific research, these bonded themes will continue to be the focal point in future decades.

Author Contributions: Conceptualization, A.M. and P.F.; writing—review and editing, A.M. and P.F.; supervision, A.M. and P.F. All authors have read and agreed to the published version of the manuscript.

Funding: This research received no external funding.

Institutional Review Board Statement: Not applicable.

Informed Consent Statement: Not applicable.

Acknowledgments: Guest editors would like to warmly thank the authors for their valuable contributions, the reviewers for their professional work and the editorial team of the journal for this collaboration.

Conflicts of Interest: The authors declare no conflict of interest.

References

1. Bonaccorsi, L.; Fotia, A.; Malara, A.; Frontera, P. Advanced adsorbent materials for waste energy recovery. *Energies* **2020**, *13*, 4299. [CrossRef]
2. Frontera, P.; Kumita, M.; Malara, A.; Nishizawa, J.; Bonaccorsi, L. Manufacturing and assessment of electrospun PVP/TEOS microfibres for adsorptive heat transformers. *Coatings* **2019**, *9*, 443. [CrossRef]
3. Frontera, P.; Macario, A.; Malara, A.; Santangelo, S.; Triolo, C.; Crea, F.; Antonucci, P. Trimetallic Ni-based catalysts over gadolinia-doped ceria for green fuel production. *Catalysts* **2018**, *8*, 435. [CrossRef]
4. Frontera, P.; Malara, A.; Modafferi, V.; Antonucci, V.; Antonucci, P.; Macario, A. Catalytic activity of Ni-Co supported metals in carbon dioxides methanation. *Can. J. Chem. Eng.* **2020**, *98*, 1924–1934. [CrossRef]
5. Malara, A.; Frontera, P.; Antonucci, P.; Macario, A. Smart recycling of carbon oxides: Current status of methanation reaction. *Curr. Opin. Green Sustain. Chem.* **2020**, *26*, 100376. [CrossRef]
6. Malara, A.; Paone, E.; Frontera, P.; Bonaccorsi, L.; Panzera, G.; Mauriello, F. Sustainable exploitation of coffee silverskin in water remediation. *Sustainability* **2018**, *10*, 3547. [CrossRef]
7. Miceli, M.; Frontera, P.; Macario, A.; Malara, A. Recovery/reuse of heterogeneous supported spent catalysts. *Catalysts* **2021**, *11*, 591. [CrossRef]
8. Nardelli, F.; Bramanti, E.; Lavacchi, A.; Pizzanelli, S.; Campanella, B.; Forte, C.; Berretti, E.; Freni, A. Thermal Stability of Ionic Liquids: Effect of Metals. *Appl. Sci.* **2022**, *12*, 1652. [CrossRef]
9. Mastronardo, E.; Piperopoulos, E.; Palamara, D.; Frazzica, A.; Calabrese, L. Morphological Observation of LiCl Deliquescence in PDMS-Based Composite Foams. *Appl. Sci.* **2022**, *12*, 1510. [CrossRef]
10. Elrashidi, A. Light Harvesting in Silicon Nanowires Solar Cells by Using Graphene Layer and Plasmonic Nanoparticles. *Appl. Sci.* **2022**, *12*, 2519. [CrossRef]
11. Bonaccorsi, L.; Donato, A.; Fotia, A.; Frontera, P.; Gnisci, A. Competitive Detection of Volatile Compounds from Food Degradation by a Zinc Oxide Sensor. *Appl. Sci.* **2022**, *12*, 2261. [CrossRef]
12. Gnisci, A.; Fotia, A.; Bonaccorsi, L.; Donato, A. Effect of Working Atmospheres on the Detection of Diacetyl by Resistive SnO_2 Sensor. *Appl. Sci.* **2022**, *12*, 367. [CrossRef]
13. Fazzino, F.; Paone, E.; Pedullà, A.; Mauriello, F.; Calabrò, P.S. A New Biorefinery Approach for the Full Valorisation of Anchovy Residues: Use of the Sludge Generated during the Extraction of Fish Oil as a Nitrogen Supplement in Anaerobic Digestion. *Appl. Sci.* **2021**, *11*, 10163. [CrossRef]
14. Satira, A.; Paone, E.; Bressi, V.; Iannazzo, D.; Marra, F.; Calabrò, P.S.; Mauriello, F.; Espro, C. Hydrothermal Carbonization as Sustainable Process for the Complete Upgrading of Orange Peel Waste into Value-Added Chemicals and Bio-Carbon Materials. *Appl. Sci.* **2021**, *11*, 10983. [CrossRef]

Article

A New Biorefinery Approach for the Full Valorisation of Anchovy Residues: Use of the Sludge Generated during the Extraction of Fish Oil as a Nitrogen Supplement in Anaerobic Digestion

Filippo Fazzino, Emilia Paone, Altea Pedullà, Francesco Mauriello and Paolo S. Calabrò *

DICEAM Department, Mediterranea University of Reggio Calabria, Via Graziella Loc. Feo di Vito, I-89122 Reggio Calabria, Italy; filippo.fazzino@unirc.it (F.F.); emilia.paone@unirc.it (E.P.); altea.pedulla@libero.it (A.P.); francesco.mauriello@unirc.it (F.M.)
* Correspondence: paolo.calabro@unirc.it

Featured Application: The biorefinery scheme proposed in this paper is almost ready for scaling up and can potentially be applied to every industrial chain involved in the preservation of fish fillets (e.g., anchovies, tuna, and salmon). This is an alternative to the current policy of making fish flour from fish remainders.

Abstract: Several anchovies species are captured all over the world; they are consumed fresh but also preserved by the industry, either by brine-fermentation or canning in oil. The industrial process generates large amounts of residue (about 50% of the original fish biomass) that is generally used to produce fish flour. In this paper, the advancement of a recently proposed process for the full valorisation of anchovies aimed at the extraction of fish oil (to be used as an omega-3 source) and at the production of biomethane through anaerobic digestion is presented. Particularly, in the experiments presented, a co-digestion of anchovy sludge—used as a nitrogen supplement—and market waste (5% and 95% on a Total Solids basis) was performed. Since the proposed extraction process uses, as a green-solvent, d-limonene, the well-known problems of toxicity for the anaerobic biomass must be overcome during the digestion process. As discussed below, the granular activated carbon (GAC) is used to reclaim and improve anaerobic digestion processes in a reactor displaying clear signs of inhibition. In fact, GAC demonstrates multiple benefits for anaerobic digestion, such as adsorption of toxic substances, biomass selection, and triggering of direct interspecies electron transfer (DIET).

Keywords: anaerobic digestion; anchovies; biorefinery; circular economy; d-limonene; granular activated carbon; inhibition

1. Introduction

In the last decade, in developed countries, the industrial practice shifted from the mere management of biomass constituted by industrial residues, by-products, and waste to their full valorisation according to biorefinery schemes. While at the turn of the millennium, in most countries, the correct landfilling of waste was considered a fully acceptable practice, nowadays the full implementation of a circular economy demands a complete valorisation of every raw material, and especially for those of biological origin [1]. A giant step in the right direction has been the adoption by the EU of the "Circular Economy Action Plan" [2] that is destined to shape, over the next decades, the European economy towards a "cleaner and more competitive future".

In many industrial chains, applied research has allowed the potential complete valorisation of several feedstocks (e.g., orange [3,4], avocado [5], and lignocellulosic biomass [6,7]).

One of the most interesting ways to implement the circular economy approach is the "blue economy", a paradigm founded on the sustainable exploitation of marine resources that allow the preservation of the marine environment, conceived as one of the key factors of global prosperity [8].

As an example of a "blue economy" process, this paper aims to demonstrate, at the laboratory level, the potential of a biorefinery scheme applied to anchovies.

Several anchovies species are captured by fisheries all over the world; according to official data [9] in the decade 2010–2019, the capture ranged between 5.5 and 11 million tonnes per year.

In addition to the direct consumption as fresh fish, anchovies are transformed either by brine-fermentation or canning (preserved in oil). These processes cause the loss, as residues (heads, viscera, and spines), of about 50% of the original fish biomass. In the conventional industrial processes, these residues are not lost but used to produce fishmeal destined for indirect human consumption (e.g., aquaculture) and, more rarely, fish oil [10,11]. Indeed, omega-3 production still strongly depends on fish oil extraction of fresh fish (especially anchovies and sardines), so much that it is competing with its aforementioned direct human consumption [12].

In this paper, the advancement of a recently proposed process [13] for the full valorisation of anchovies aimed at the extraction of fish oil (to be used as an omega-3 source) by their residues and the production of biomethane through anaerobic digestion is presented (Figure 1). The produced biomethane can be converted to energy by very efficient combined heat and power schemes, allowing not only to cover the energy demand of the process itself but also to sell surplus energy.

The proposed extraction process uses, as a green-solvent, d-limonene, namely a widely available by-product of the citrus industry [3]. As a by-product, an anchovy sludge (AS) is obtained in considerable amount (97% w/w of the total residue quantity) [10]. AS is expected to be rich in d-limonene and poor in nutrient elements after oil extraction and, thus, unsuitable for animal feeding or other similar forms of valorisation.

The advancement proposed and presented in this paper is the evaluation of the potential of by-products derived from fish oil extraction (AS), to produce biomethane by anaerobic digestion, both by batch and by semi-continuous experiments without any previous pre-treatment. In particular, semi-continuous experiments afford for a preliminary optimization of the process, solving, at least partially, two of the problems linked to the use of AS a substrate for anaerobic digestion: the high presence of d-limonene (whose inhibition capacity during anaerobic digestion is well known [14,15]) and the unbalanced C/N ratio.

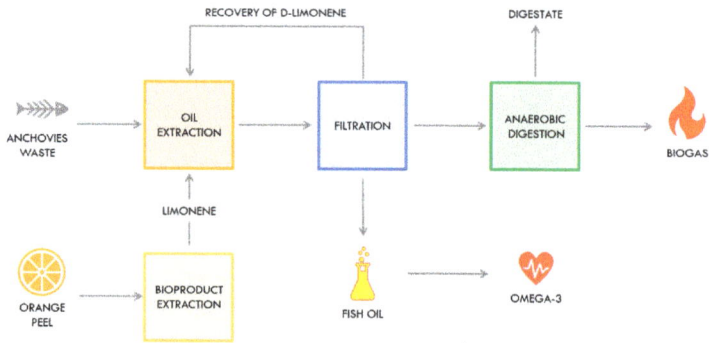

Figure 1. Biorefinery scheme for the full valorisation of anchovy waste.

As discussed below, granular activated carbon (GAC) was used to reclaim and improve the anaerobic digestion processes in a reactor showing clear signs of inhibition. In fact,

GAC demonstrated multiple benefits for anaerobic digestion, such as adsorption of toxic substances, triggering direct interspecies electron transfer (DIET) [15–21].

2. Materials and Methods

2.1. Fish Oil Extraction Process

After the homogenization of anchovy residues (300 g) by a blender is carried out, they are mixed with a first aliquot of d-limonene (150 g) refrigerated at 4 °C. The so-obtained semi-solid slurry is added to a second aliquot of cold d-limonene (150 g) into a glass beaker sealed with aluminium foil and further coated with parafilm. The mixture is then magnetically stirred at 700 rpm for 24 h at room temperature, with the fish oil obtained by rotavaporing the supernatant at 90 °C (pressure: 40 mbar). The by-product of the process is anchovy sludge (AS), still rich in d-limonene.

2.2. Substrates and Inoculum Characterisation

The main substrate used in the experiments was the AS derived from fish oil extraction. A co-substrate mimicking fruit and vegetable market waste (MW), and composed of 49.0% w/w of potatoes, 44.4% w/w of apples, and 6.6% w/w of carrots, was also used. Both substrates were characterized by measuring the total and volatile solids, pH, and carbon to nitrogen ratio (C/N), while d-limonene content was determined only in the AS. Total and volatile solids and pH were evaluated according to standard methods [22], the C/N was measured using an elemental analyser TOC-LCSH (Shimadzu; Kyoto, Japan) while the analysis of the residual d-limonene, present in the substrate, was carried out according to the analytical procedure previously reported [15], by mixing 0.3 g of AS with 3 mL of a toluene solution (as an internal standard) in cyclohexane (0.1 M) for 6 h. The liquid suspension was then filtered and injected into an offline GC-FID (Agilent 6890 N) equipped with a CP-WAX 52CB column (60 m, i.d. 0.53 mm).

The inoculum used in batch and semi-continuous experiments was a digestate coming from previous experiments. It was characterized by measuring total and volatile solids and pH evaluated according to standard methods [22]. Table 1 reports the main characteristics of substrates and inocula.

Table 1. Substrates and inocula characteristics.

	Anchovy Sludge	Market Waste	Inoculum (Batch Tests)	Inoculum (Semi-Continuous t.)
TS [%]	20.1	19.4	3.9	3.1
VS [%TS]	66.7	93.3	66.6	66.7
pH	6.85	5.26	8.13	8.04
C/N	3.4	36.3	-	-
d-limonene [g/g]	0.125 [1] 0.160 [2]	-	-	-

[1] A sample of AS used for batch and semi-continuous tests (days 1–56); [2] A sample of AS used for semi-continuous tests (days 57–80).

2.3. Biomethane Potential (BMP) Tests

BMP tests (Table 2) were performed in triplicate under mesophilic conditions using a self-developed method [23,24], basically compliant with UNI/TS 11703:2018 (the Italian standard procedure for BMP tests). Tests were performed using glass bottles (1.1 L volume) placed in a thermostatic cabinet at 35 ± 0.5 °C and mixed by using a magnetic stirrer. The inoculum was mixed with the substrate (substrate to inoculum ratio in terms of vs. was set equal to 0.3) and nutrient solutions (prepared and dosed according to UNI/TS 11703:2018).

BMP tests were performed to evaluate the potential methane production from AS, MW, and from a mixture of both. The mixture was prepared with the aim to obtain a C/N in the substrate equal to 25, which can be considered a well-balanced value. To obtain the desired C/N, a proportion of 1:19 (5–95) on the TS basis of AS and MW, respectively, was necessary.

Table 2. BMP design of experiments.

Substrate	Market Waste (MW)	Anchovy Sludge (AS)	Mix (95% MW + 5% AS)
pH	8.1	8.1	8.1
C/N	36.31	3.41	24.73
$gVS_{substrate}/gVS_{inoculum}$	0.30	0.30	0.30
TS [g]	3.35	4.69	3.40
TS at the beginning of the experiment	3.17%	3.39%	3.18%

In addition to BMP bottles, blanks (containing inoculum and a nutrients solution, used to assess a non-specific biomethane production) and cellulose-fed reactors (used as control) were also prepared. About three times per week, the biogas produced was transferred in a bottle filled with a NaOH solution (3M) for CO_2 adsorption, and the methane amount in the produced biogas was then measured by an eudiometer (a water displacement method). The pH was measured at the beginning of the tests, while TS, VS, COD, ammonium ion and chloride concentration, VFAs, and FOS/TAC were measured at the end of them. VFAs and FOS/TAC allow verification of the stability of the digestion process since, when a high level of them is registered or when they tend to increase over time, an unbalance of the process, due to an overloading or an inhibition of the methanogenesis, is possible [25,26]. TS, VS, and pH were measured using standard methods [22]; COD, ammonium ion, and chloride concentration were evaluated thanks to a photometric method (Photometer WTW Photolab S12 and appropriate pre-dosed cuvettes), whereas VFAs were determined through a three-point titration method, and then the FOS/TAC was calculated [25,26].

2.4. Semi-Continuous Experiments

Semi-continuous experiments were carried out with the aim to reproduce more precisely, at a laboratory scale, the digestion process. A Bioprocess Control Bioreactor simulator system equipped with 4 continuously stirred tank reactors (glass, working volume 1.8 L) placed in a thermostatic water bath (set at an operating temperature of 35 °C) was used. This system allows the feeding and discharge of the reactors and the measurement of the produced biomethane by a patented system based on water/gas displacement. The hydraulic retention time was set equal to 20 days, while the organic loading rate, initially set at 2.0 $gVS·L^{-1}·day^{-1}$ during the start-up (days 0–38), was reduced before the beginning of the regime phase (days 39–83—more than 2·HRT), since a severe overloading was evident in all the reactors. In order to accelerate the recovery of the reactors, the supplementation of new inoculum was also necessary in some experiments and, for this reason, only data recorded in the regime phase are presented and discussed. The reactors were fed three times per week; the pH was measured during each feeding/discharge and $NaHCO_3$ was added if the measure value was <6.7.

A composite weekly sample was prepared for analyses of TS, VS, COD, ammonium ion concentration, VFAs, and FOS/TAC using the same methods mentioned for batch tests. During the operation, due to the change of the AS and to the subsequent increase of the *d*-limonene, signs of inhibition of the process were evident. For this reason, as already mentioned, 10 $g·L^{-1}$ of granular activated carbon (CARBOSORB 2040, 20 × 40 mesh; Comelt srl, Milan, Italy) were added in reactor 3; this concentration was then kept constant until the end of the test. The test was stopped after 83 days due to the unavailability of the lab for the following weeks due to reasons independent of the will of the authors. Table 3 summarizes the main characteristics of the reactors during the regime phase (days 39–83).

Table 3. Semi-continuous experiments.

	Reactor 1	Reactor 2	Reactor 3	Reactor 4
Reinoculation (end of start-up phase)	YES	YES	NO	YES
Loading (regime phase) [$gVS \cdot L^{-1} \cdot day^{-1}$]	1	0.5	1	0.5
Market Waste (TS basis)	100%	100%	95%	95%
Anchovy Sludge (TS basis)	-	-	5%	5%
C/N substrate	36.3	36.3	24.7	24.7
Substrate addition [$g \cdot d^{-1}$] (regime phase)	10.00	5.00	10.10	5.05
Expected regime d-limonene conc. [$mg \cdot L^{-1}$]	-	-	680 [1] 870 [2]	340 [1] 436 [2]
Addition of GAC—10 $g \cdot L^{-1}$ (days)	-	-	74–83	-

[1] Anchovy Sludge 1—days 1–56; [2] Anchovy Sludge 2—days 57–83.

3. Results and Discussion

3.1. BMP Tests

Methane production during the BMP test was regular for batches fed with the MW or the mixture between the former and AS (Figure 2 and Table 4). Final BMP values were very similar: 421 ± 13 $NmL \cdot gVS^{-1}$ for MW and 420 ± 23 $NmL \cdot gVS^{-1}$ for the mixture respectively; the very low standard deviation witnesses the uniformity of the production among batches. Moreover, results confirm those reported in the scientific literature for similar substrates [27,28], corroborating that MW is an excellent substrate for anaerobic digestion. Notably, due to the supply of nitrogen from the inoculum, the benefit of optimizing the C/N ration by adding the AS is not evident in batch tests.

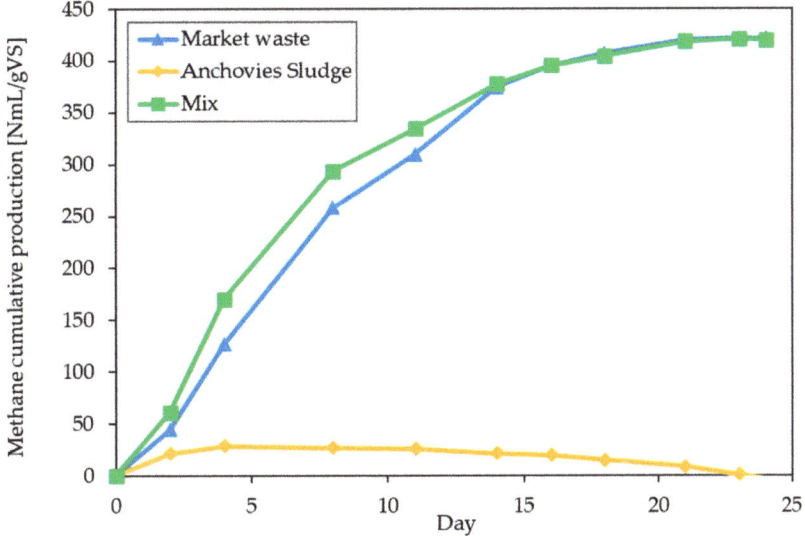

Figure 2. Methane cumulative production during BMP tests.

The results relative to batches fed uniquely with AS are completely different, even if the difference was expected. At the end of the experiment, their production was, on average, lower than that of the blank (inoculum); in this case, the standard deviation was also very limited, thus confirming the consistency of the results. The most likely reason for the observed behaviour is due to the presence of d-limonene in the substrate. In fact, the initial concentration of the latter at the beginning of the experiment was of about 4870 $mg \cdot L^{-1}$ and therefore well above the level severely inhibiting anaerobic digestion [14].

The analyses on digestate at the end of the experiments (Table 4) further confirm these trends: while the reactors fed with MW and with the mixture perform similarly (and similarly to the positive control fed with cellulose data; not shown) in terms of residual COD, ammonium ion and chloride concentration, and VFA, the batches fed with AS only present a higher residual COD (+63%) and a high concentration of VFA (+510%). The latter indicates that the conversion of the substrate to VFA occurs, but the high concentration of d-limonene severely affects the methanogens, triggering the accumulation of VFA up to toxic values [29–31]. The results of the batch tests indicate that AS, even in the presence of quite high amounts of d-limonene, is potentially a good substrate for co-digestion with carbonaceous feedstocks.

Table 4. Analyses of digestate from BMP experiments.

	Market Waste	Anchovy Sludge	Mix (Market w. + Anch. Sludge)
pH	7.6 ± 0.00	7.6 ± 0.06	7.5 ± 0.01
COD [mg/L]	7008 ± 398	11470 ± 130	7073 ± 385
ammonium ion [mg/L]	1435 ± 60	1861 ± 61	1411 ± 39
chloride [mg/L]	1280 ± 93	1563 ± 105	1363 ± 274
VFA [mg/L]	550 ± 156	3662 ± 69	651 ± 129
FOS/TAC	0.11 ± 0.03	0.4 ± 0.02	0.12 ± 0.02

3.2. Semi-Continuous Experiments

In the initial part of the regime phase (Figure 3b), methane production seems to be linked only to the applied organic loading. In fact, reactors 1 and 3, and 2 and 4, respectively, behave similarly. This behaviour was confirmed for reactors 2 and 4 (low loading) until the end of the experiment, while reactors 1 and 3 present a different production pattern. Reactor 1 and 3's productions slowed gradually since about day 60 for the latter; this tendency was more pronounced, but a sudden recovery was also evident from about day 75 of the total operation.

For the first 20 days of regime phase (1·HRT) the average yield (Figure 3b) was similar for reactors 1 and 3, and 2 and 4, respectively; it was equal to about 0.2 NL·gVS$_{added}^{-1}$ for reactors 1 and 3 and to about 0.25 NL·gVS$_{added}^{-1}$ for reactors 2 and 4, respectively. Then a continuous decrease was evident for reactor 1 that reached a value of about 0.17 NL·gVS$_{added}^{-1}$ in the last days of the experiment. On the contrary, after a sharper decrease until day 75, first stabilization and then a slight increase was registered; in the last days of operation the yield of reactor 3 surpassed that of reactor 1.

pH (Figure 3c) in all reactors during the regime phase was close to 7, seldom needing NaHCO$_3$ addition to increase the buffering capacity. A marked tendency toward a reduction is evident in reactor 3 after the beginning of the feeding of the new sludge.

TS (data not displayed) and VS concentrations (Figure 3d) were stable during the regime phase; TS were slightly higher for reactors 1 and 3 (loading 1 gVS·L^{-1}·day^{-1}).

The COD concentration (Figure 3e), except for a few spikes (e.g., week 9—reactor 2), is quite low when the process is stable and tends to increase when the process is inhibited (reactors 1 and 4, since weeks 9–10).

The ammonium ion concentration (Figure 3f) displays a tendency to decrease, and reaches very low values (it is practically absent) in reactor 1 since week 4 of operation and in reactor 2 since week 6.

The VFA concentration and FOS/TAC (Figure 3g,h) are two important process stability indicators, they are conveniently low for reactors 2 and 4, while they display a tendency to increase since week 10 for reactors 1 and 3. In reactor 3, after the beginning of GAC addition, VFA suddenly decreases and the FOS/TAC stabilizes.

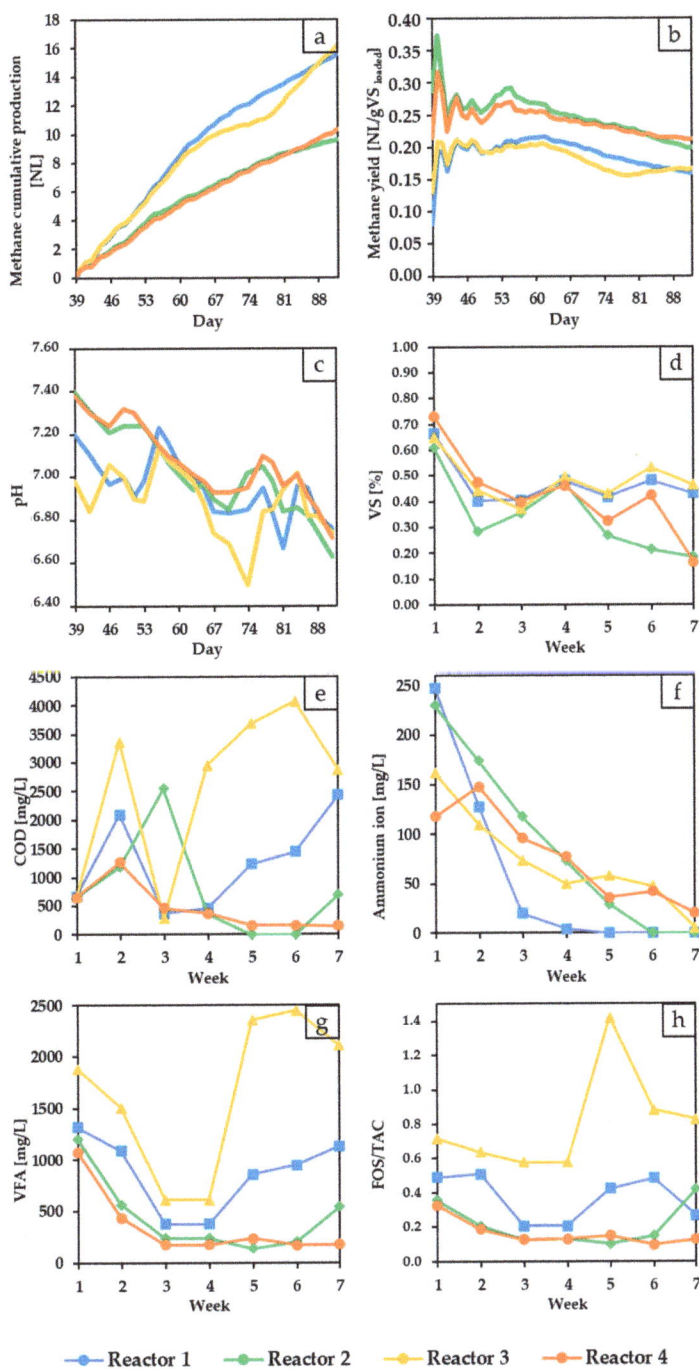

Figure 3. Semi-continuous experiments results: (**a**) methane cumulative production, (**b**) methane yield, (**c**) pH, (**d**) volatile solids, (**e**) COD, (**f**) ammonium ion, (**g**) VFA, and (**h**) FOS/TAC.

It is worth pointing out that the reactors during the start-up phase, as mentioned above, suffered from a severe overloading that, however, did not seem to influence significantly the regime phase, indicating a good recovery. The fact that differences among reactors operating at the same loading, although using different substrates (either MW only or a mix between this and AS), were not detected during the initial regime phase can be attributed to two main factors: (i) the supply of nitrogen in reactors 1 and 2 (fed only with MW) was linked to reinoculation (see Table 2), operating to re-establish the process after the already mentioned overloading, and (ii) the adaptation to d-limonene during the start-up phase of the microbial population of reactor 3 (the one with the highest d-limonene loading, as reported in Table 2) was the only one not needing reinoculation.

The reasons for the reduction of the methane production after about 1·HRT in the regime phase in reactors 1 and 3 can be attributed to two different factors. For reactor 1, the reduction of nitrogen below tolerable limits is evident (see Figure 3f) and this supports the idea that C/N optimization is essential for stable anaerobic digestion. Indeed, adequate nitrogen presence in digesters must be ensured since it is involved in the fundamental activities of microbial metabolism (synthesis of proteins, enzymes, ribonucleic acid (RNA), and deoxyribonucleic (DNA)) [32]. Thus, the lack of nitrogen could have affected the anaerobic bacteria's metabolism, eventually causing process failure.

For reactor 3, besides the nitrogen reduction also detected in this case, the feeding of the new AS (see Table 2) most probably increases the d-limonene concentration up to intolerable levels and triggers an inhibition of the methanogenesis, as the sharp increase in VFA concentration and in FOS/TAC and the significant pH reduction both witness. The supplementation of GAC (10 g·L^{-1}) since day 74 and until the end of the experiment causes an almost immediate recovery of the reactor, with a sharp increase in methane production since day 77 and until the end of the experiment. These results confirm previous research [15,21] on the potential of this material in sustaining the anaerobic digestion of d-limonene containing substrates. Also, if the experiment, as already mentioned, was forcedly terminated after about 10 days since the beginning of the GAC addition, its effect on the process stabilization (probably mainly linked to the adsorption of d-limonene) are evident. It is interesting to note that in reactor 4, with a potential d-limonene concentration close to 450 mg·L^{-1}, the process does not demonstrate signs of disruption, confirming the potential of the microbial community to adapt to d-limonene and therefore the importance of an optimized loading when using AS for the co-digestion with a carbonaceous substrate. This optimization should aim to slowly increase the quantity fed to the reactor to keep the d-limonene concentration at a tolerable level. On the other hand, two other factors are very important and worth noting: (i) optimization of the recovery of d-limonene during the extraction of the fish oil would be beneficial for the entire biorefinery scheme, and (ii) GAC is confirmed as a powerful additive in the anaerobic digestion of substrates containing d-limonene.

The yield of the process was lower than expected, compared to the BMP value for the co-digestion of MW and AS (420 ± 23 NmL·gVS^{-1}) with respect to batch tests; the reduction is evident and in the order of 40% for the low loaded reactors and 50% for the others. Furthermore, this reaction is more pronounced than the 10–30% often reported in the scientific literature for batch and semi-continuous tests on the same substrate [33–35]. Since only reactors also fed with MW display similar behaviour, this situation is most probably attributable more to the imperfect start-up of the reactors than to the use of AS as a co-substrate.

4. Conclusions

This paper demonstrates the potential suitability of AS a co-substrate for the anaerobic digestion of mainly carbonaceous feedstocks. However, the presence of d-limonene is an issue that requires proper countermeasures; the first is the optimization of the oil extraction process to reduce the residual of the solvent present in AS. In addition, the results presented here, although preliminary, demonstrate how proper adaptation and the supplementation of GAC during anaerobic digestion can improve tolerance to *d*-limonene.

Author Contributions: Conceptualization, F.M. and P.S.C.; methodology, F.F., F.M., E.P., A.P. and P.S.C.; investigation, F.F., E.P. and A.P.; resources, F.M. and P.S.C.; writing—original draft preparation, F.F., F.M., E.P., A.P. and P.S.C.; writing—review and editing, F.F., F.M., E.P., A.P. and P.S.C.; visualization, F.F., E.P. and A.P.; supervision, F.M. and P.S.C. All authors have read and agreed to the published version of the manuscript.

Funding: This research received no external funding.

Data Availability Statement: The data presented in this study are available from the corresponding author upon request. The data are not publicly available due to the fact that the research is still ongoing and a patent could be requested.

Acknowledgments: The authors sincerely thank Daniela Pizzone for the preparation of anchovy residues. This work is dedicated to R. Pietropaolo, in honour of his 80th birthday.

Conflicts of Interest: The authors declare no conflict of interest.

References

1. Ellen MacArthur Foundation Universal Circular Economy Policy Goals: Enabling the Transition to Scale. Available online: https://www.ellenmacarthurfoundation.org/publications/universal-circular-economy-policy-goals-enabling-the-transition-to-scale (accessed on 27 July 2021).
2. European Commission. *Circular Economy Action Plan*; European Commission: Brussels, Belgium, 2020; p. 28. [CrossRef]
3. Zema, D.A.; Calabrò, P.S.; Folino, A.; Tamburino, V.; Zappia, G.; Zimbone, S.M. Valorisation of citrus processing waste: A review. *Waste Manag.* **2018**, *80*, 252–273. [CrossRef]
4. Fazzino, F.; Mauriello, F.; Paone, E.; Sidari, R.; Calabrò, P.S. Integral valorization of orange peel waste through optimized ensiling: Lactic acid and bioethanol production. *Chemosphere* **2021**, *271*, 129602. [CrossRef] [PubMed]
5. García-Vargas, M.C.; Contreras, M.d.M.; Castro, E. Avocado-Derived Biomass as a Source of Bioenergy and Bioproducts. *Appl. Sci.* **2020**, *10*, 8195. [CrossRef]
6. Câmara-Salim, I.; Conde, P.; Feijoo, G.; Moreira, M.T. The use of maize stover and sugar beet pulp as feedstocks in industrial fermentation plants—An economic and environmental perspective. *Clean. Environ. Syst.* **2021**, *2*, 100005. [CrossRef]
7. Rajesh Banu, J.; Preethi; Kavitha, S.; Tyagi, V.K.; Gunasekaran, M.; Karthikeyan, O.P.; Kumar, G. Lignocellulosic biomass based biorefinery: A successful platform towards circular bioeconomy. *Fuel* **2021**, *302*, 121086. [CrossRef]
8. The Blue Economy—Home. Available online: https://www.theblueeconomy.org/ (accessed on 27 July 2021).
9. FAO. FAO Fisheries & Aquaculture—Fishery Statistical Collections—Global Capture Production. Available online: https://www.fao.org/fishery/statistics/global-capture-production/en (accessed on 24 October 2021).
10. Laso, J.; Margallo, M.; Serrano, M.; Vázquez-Rowe, I.; Avadí, A.; Fullana, P.; Bala, A.; Gazulla, C.; Irabien, Á.; Aldaco, R. Introducing the Green Protein Footprint method as an understandable measure of the environmental cost of anchovy consumption. *Sci. Total Environ.* **2018**, *621*, 40–53. [CrossRef]
11. Laso, J.; Margallo, M.; Celaya, J.; Fullana, P.; Bala, A.; Gazulla, C.; Irabien, A.; Aldaco, R. Waste management under a life cycle approach as a tool for a circular economy in the canned anchovy industry. *Waste Manag. Res.* **2016**, *34*, 724–733. [CrossRef]
12. Shepherd, C.J.; Jackson, A.J. Global fishmeal and fish-oil supply: Inputs, outputs and marketsa. *J. Fish. Biol.* **2013**, *83*, 1046–1066. [CrossRef]
13. Paone, E.; Fazzino, F.; Pizzone, D.M.; Scurria, A.; Pagliaro, M.; Ciriminna, R.; Calabrò, P.S. Towards the anchovy biorefinery: Biogas production from anchovy processing waste after fish oil extraction with biobased Limonene. *Sustainability* **2021**, *13*, 2428. [CrossRef]
14. Ruiz, B.; Flotats, X. Citrus essential oils and their influence on the anaerobic digestion process: An overview. *Waste Manag.* **2014**, *34*, 2063–2079. [CrossRef] [PubMed]
15. Calabrò, P.S.; Fazzino, F.; Folino, A.; Paone, E.; Komilis, D. Semi-Continuous Anaerobic Digestion of Orange Peel Waste: Effect of Activated Carbon Addition and Alkaline Pretreatment on the Process. *Sustainability* **2019**, *11*, 3386. [CrossRef]
16. Wu, Y.; Wang, S.; Liang, D.; Li, N. Conductive materials in anaerobic digestion: From mechanism to application. *Bioresour. Technol.* **2020**, *298*, 122403. [CrossRef]
17. Pan, C.; Fu, X.; Lu, W.; Ye, R.; Guo, H.; Wang, H.; Chusov, A. Effects of conductive carbon materials on dry anaerobic digestion of sewage sludge: Process and mechanism. *J. Hazard. Mater.* **2020**, *384*, 121339. [CrossRef] [PubMed]
18. Baek, G.; Kim, J.J.; Kim, J.J.; Lee, C. Role and potential of direct interspecies electron transfer in anaerobic digestion. *Energies* **2018**, *11*, 107. [CrossRef]
19. Yang, Y.; Zhang, Y.; Li, Z.; Zhao, Z.; Quan, X.; Zhao, Z. Adding granular activated carbon into anaerobic sludge digestion to promote methane production and sludge decomposition. *J. Clean. Prod.* **2017**, *149*, 1101–1108. [CrossRef]

20. Capaccioni, B.; Caramiello, C.; Tatàno, F.; Viscione, A. Effects of a temporary HDPE cover on landfill gas emissions: Multiyear evaluation with the static chamber approach at an Italian landfill. *Waste Manag.* **2011**, *31*, 956–965. [CrossRef] [PubMed]
21. Calabrò, P.S.; Fazzino, F.; Folino, A.; Scibetta, S.; Sidari, R. Improvement of semi-continuous anaerobic digestion of pre-treated orange peel waste by the combined use of zero valent iron and granular activated carbon. *Biomass Bioenergy* **2019**, *129*, 105337. [CrossRef]
22. APHA; AWWA; WEF. *Standard Methods for the Examination of Water and Wastewater*, 22nd ed.; Rice, E.W., Baird, R.B., Eaton, A.D., Clesceri, L.S., Eds.; American Public Health Association, American Water Works Association, Water Environment Federation: Washington, DC, USA, 2012; ISBN 9780875530130.
23. Calabro, P.S.; Panzera, M.F. Biomethane production tests on ensiled orange peel waste. *Int. J. Heat Technol.* **2017**, *35*, S130–S136. [CrossRef]
24. Calabro', P.S.; Folino, A.; Fazzino, F.; Komilis, D. Preliminary evaluation of the anaerobic biodegradability of three biobased materials used for the production of disposable plastics. *J. Hazard. Mater.* **2020**, *390*, 121653. [CrossRef]
25. Buchauer, K. A comparison of two simple titration procedures to determine volatile fatty acids in influents to waste-water and sludge treatment processes. *Water SA* **1998**, *24*, 49–56.
26. Liebetrau, J.; Pfeiffer, D.; Thrän, D. *Collection of Methods for Biogas*; Federal Ministry for Economic Affairs and Energy (BMWi): Berlin, Germany, 2016.
27. Lopez, V.M.; De la Cruz, F.B.; Barlaz, M.A. Chemical composition and methane potential of commercial food wastes. *Waste Manag.* **2016**, *56*, 477–490. [CrossRef]
28. Masebinu, S.O.; Akinlabi, E.T.; Muzenda, E.; Aboyade, A.O.; Mbohwa, C. Experimental and feasibility assessment of biogas production by anaerobic digestion of fruit and vegetable waste from Joburg Market. *Waste Manag.* **2018**, *75*, 236–250. [CrossRef]
29. Forgács, G.; Pourbafrani, M.; Niklasson, C.; Taherzadeh, M.J.; Hováth, I.S.; Forgcs, G.; Pourbafrani, M.; Niklasson, C.; Taherzadeh, M.J.; Hováth, I.S. Methane production from citrus wastes: Process development and cost estimation. *J. Chem. Technol. Biotechnol.* **2012**, *87*, 250–255. [CrossRef]
30. Wikandari, R.; Millati, R.; Cahyanto, M.N.; Taherzadeh, M.J. Biogas Production from Citrus Waste by Membrane Bioreactor. *Membrane* **2014**, *4*, 596–607. [CrossRef] [PubMed]
31. Wang, Y.; Zhang, Y.; Wang, J.; Meng, L. Effects of volatile fatty acid concentrations on methane yield and methanogenic bacteria. *Biomass Bioenergy* **2009**, *33*, 848–853. [CrossRef]
32. Parkin, G.F.; Owen, W.F. Fundamentals of Anaerobic Digestion of Wastewater Sludges. *J. Environ. Eng.* **1986**, *112*, 867–920. [CrossRef]
33. Zhang, C.; Su, H.; Tan, T. Batch and semi-continuous anaerobic digestion of food waste in a dual solid-liquid system. *Bioresour. Technol.* **2013**, *145*, 10–16. [CrossRef]
34. Ruffino, B.; Fiore, S.; Roati, C.; Campo, G.; Novarino, D.; Zanetti, M. Scale effect of anaerobic digestion tests in fed-batch and semi-continuous mode for the technical and economic feasibility of a full scale digester. *Bioresour. Technol.* **2015**, *182*, 302–313. [CrossRef] [PubMed]
35. Browne, J.D.; Allen, E.; Murphy, J.D. Assessing the variability in biomethane production from the organic fraction of municipal solid waste in batch and continuous operation. *Appl. Energy* **2014**, *128*, 307–314. [CrossRef]

Article

Hydrothermal Carbonization as Sustainable Process for the Complete Upgrading of Orange Peel Waste into Value-Added Chemicals and Bio-Carbon Materials

Antonella Satira [1,2], Emilia Paone [1,3,*], Viviana Bressi [2], Daniela Iannazzo [2], Federica Marra [4], Paolo Salvatore Calabrò [1], Francesco Mauriello [1] and Claudia Espro [2,*]

[1] Dipartimento DICEAM, Università Mediterranea di Reggio Calabria, Loc. Feo di Vito, 89122 Reggio Calabria, Italy; antonella.satira@unirc.it (A.S.); paolo.calabro@unirc.it (P.S.C.); francesco.mauriello@unirc.it (F.M.)
[2] Dipartimento di Ingegneria, Università di Messina, Contrada di Dio—Vill. S. Agata, 98166 Messina, Italy; viviana.bressi@unime.it (V.B.); diannazzo@unime.it (D.I.)
[3] Consorzio Interuniversitario per la Scienza e la Tecnologia dei Materiali (INSTM), 50121 Firenze, Italy
[4] Dipartimento di Agraria, Università Mediterranea di Reggio Calabria, Loc. Feo di Vito, 89122 Reggio Calabria, Italy; fede.marra6@gmail.com
* Correspondence: emilia.paone@unirc.it (E.P.); espro@unime.it (C.E.)

Featured Application: Hydrothermal carbonization process can be efficiently used for the simultaneous production of value-added chemicals, including furans and levulinates, and carbon-based materials.

Abstract: In this study, a simple and green protocol to obtain hydrochar and high-added value products, mainly 5-hydroxymethylfurfural (5-HMF), furfural (FU), levulinic acid (LA) and alkyl levulinates, by using the hydrothermal carbonization (HTC) of orange peel waste (OPW) is presented. Process variables, such as reaction temperature (180–300 °C), reaction time (60–300 min), biomass:water ratio and initial pH were investigated in order to find the optimum conditions that maximize both the yields of solid hydrochar and 5-HMF and levulinates in the bio-oil. Data obtained evidence that the highest yield of hydrochar is obtained at a 210 °C reaction temperature, 180 min residence time, 6/1 w/w orange peel waste to water ratio and a 3.6 initial pH. The bio-products distribution strongly depends on the applied reaction conditions. Overall, 180 °C was found to be the best reaction temperature that maximizes the production of furfural and 5-HMF in the presence of pure water as a reaction medium.

Keywords: orange peel waste (OPW); hydrothermal carbonization; hydrochar; 5-hydroxymethylfurfural (5-HMF); furfural (FU); levulinic acid (LA)

1. Introduction

Citrus fruits are among the most cultivated and processed fruits worldwide, with an annual production of about 152 million tons [1]. The citrus processing industry generates huge amounts of residues mainly in the form of pulp and peels, with the latter accounting for almost 50% of the wet fruit mass [2]. With more than 50 million metric tons in 2020 [3], the juice industry alone generates a huge volume of orange peel waste (OPW) that requires suitable management, taking into consideration the high OPW biodegradability that causes its fast fermentation, which is often uncontrolled. Therefore, the direct disposal of this secondary product without previous proper processing, raises serious environmental issues and economic loss for the citrus industry since traditional disposal strategies, such as incineration or landfilling, are expensive and insufficient in terms of environmental protection and energy efficiency [4,5].

To date, several technological innovations have been proposed to manage the OPW, mainly aimed at converting the bio-waste into a valuable resource for the extraction of essential oil and the sustainable production of biogas, bioethanol and other biobased chemicals such as pectin, bioethanol and acids [6–9].

Moreover, innovative approaches, such as the catalytic upgrading and the extraction of bioactive compounds, leading to an ecofriendly production of active ingredients having applications in different sectors, have been proposed in the last years. Even if the valorization of agro-industrial wastes by the preparation of bioactive materials can be considered a sustainable methodology, other factors also need to be taken into account, including the extraction method, the employed reagents and solvents, general expense and employing the use of toxic chemicals with long time/energy-demanding procedures, poor selectivity, and large volumes of solvents [10]. Furthermore, considering that citrus processing wastes contain high amounts of moisture, there is a great interest in their use as feedstock for thermochemical processes as well [11]. The hydrothermal carbonization (HTC) is simply a thermochemical conversion process carried out in a water medium under autogeneous pressure at a relatively mild temperature (180–300 °C) and represents a promising treatment technique for the wet lignocellulosic biomass waste since it permits overcoming the drawbacks of conventional thermochemical processes that require the use of dry feedstocks [11]. During the hydrothermal carbonization, biomass is dehydrated in-situ and processed into solid, liquid and non-condensable gaseous products. The distribution of products depends on the process conditions and on the feedstock used, but in general 50–80% of the original biomass is present in the solid product, 5–20% in the aqueous phase which contains inorganic and organic matter, and 2–5% in the gas phase which is mainly composed of CO_2 [12]. Hydrothermal methods are largely used in petroleum-based refineries and are getting more and more attention from both scientific and industrial researchers since their process parameters can be easily translated into modern biorefineries aimed at the upgrading of lignocellulosic residues and wastes [13]. With respect to other (bio)refinery processes, HT protocols present several advantages in terms of sustainability since they generally adopt mild reaction conditions (e.g., temperature and pressure) without any homogeneous or heterogeneous catalysts and in the presence of water as a green reaction solvent (used as such or in combination with simple aliphatic alcohols) [13].

The carbonaceous residue obtained by the HTC of the citrus processing waste is rich in oxygenated functional groups, making it a promising material in a wide range of applications, including pollutants adsorption [14], soil amendment [15], as fuel in energy applications [11] and as a low-cost material for capacitor and sensing applications, as recently reported by some of the authors [16–19].

Recently, bio carbon-based catalysts, obtained via the hydrothermal carbonization of the orange peel, have been also successfully adopted for acid treatments of the lignocellulosic biomass aimed at biodiesel production [20], together with platform molecules such as xylose, levulinic acid and its derivatives [21,22]. Aqueous and gas phases obtained during the HTC of orange peels and other citrus fruit wastes are generally considered as by-products and only a few studies on their use as resources of added value chemicals have been reported [23]. In this regard, it should be emphasized that the light bio-oil, obtained by extraction from the aqueous phase resulting from the hydrothermal carbonization of citrus wastes, consists mainly of aldehydes, phenols, ketones, acids, and some small molecules and heterocyclic compounds of potential interest, such as feedstocks for the synthesis of chemicals and liquid bio-fuels [24]. Therefore, the optimization of the process operating parameters of the HTC, to obtain simultaneously biocarbon materials and value-added products, could represent a valid approach aimed at minimizing the environmental and economic impact caused by the management of the agro-industrial waste. Among the considerable range of chemical building blocks that can be obtainable from citrus wastes, furfural (FU) and 5-hydroxymethylfurfural (HMF) have shown a great potential in replacing fossil-derived molecules in the synthesis of valuable chemicals—including levulinate derivatives—and pharmaceuticals [25].

In this paper, the effect of process variables on the hydrothermal carbonization (HTC) of the orange peel as industrial processing waste is investigated for both hydrochar and furan derivative (FU and 5-HMF) in the bio-oil liquid fraction, in order to optimize the yields to solid and liquid fractions. The exploration of the single-step hydrothermal treatment represents a promising example of the wet organic waste valorization to produce value-added products with high yields and, at the same time, to avoid potential and serious environmental issues arising from the citrus processing waste management and disposal. Particular attention was given to the peculiar chemical composition of the liquid fraction obtained from the HT process that is rich in components of absolute strategic interest for the chemical and pharmaceutical industry.

2. Materials and Methods

2.1. Raw Materials

The orange peels (OPW), obtained from an industry located in Sicily (Italy), were grounded to a particle size smaller than 2 mm and stored in a sealed plastic bag at $-20\ °C$ (in order to decrease the rate of degradation and the loss of volatile matter) and defrosted before their use in the experimental tests.

Acetic acid (CH_3COOH) $\geq$ 99.8 $w/w\%$ and sulfuric acid (H_2SO_4) 95.0–97.0 $w/w\%$ as well as other chemicals were purchased from Merk Life Science S.r.l.

2.2. HTC Experimental Procedure

In a typical run, a mixture of wet OPW and deionized water at the desired biomass to water ratio was ultrasonically agitated for 15 min at room temperature (20 °C) and then transferred into a 300 mL stainless steel autoclave (series 4540 Parr Instrument Company, IL, USA) for HTC. Since the water to solid ratio has a significant effect on the reaction products, a series of experiments setting the L:S ration to 4:1, 6:1, 12:1 and 24:1 (w/w) were performed. For all the experiments, a fixed amount of water was used. Prior to reaction, residual air was removed from the sealed reactor by repeatedly pressurizing with nitrogen and venting to atmosphere. In a typical HTC test, the reaction mixture was heated under autogenous pressure up to the reaction temperature (150–300 °C), at a heating rate of 5 °C min^{-1} continuously monitored through a thermocouple placed into the autoclave and connected to the reactor controller within the whole. The residence time, after reaching the reaction temperature, was set at 30 min, 60 min, 180 min and 300 min, at a stirring speed of 600 rpm. After the HTC reaction, the autoclave was rapidly cooled at room temperature. Then, the HTC solid and liquid products were separated by vacuum filtration, with a Buchner funnel and filter paper. Afterwards, 40 mL of the obtained liquid aqueous product was extracted by 40 mL of diethyl ether in a separation funnel for carrying out the GC-MS analysis. A rotary evaporator was used for removing diethyl ether at 40 °C. Each extraction was performed twice (150 mL of diethyl ether). The remaining liquid was defined as light bio-oil in this study. Anhydrous sodium sulfate was used as a drying agent of the light bio-oil after the extraction. The obtained light bio-oil samples were hereafter named as L-HC T-t, where T denotes the reaction temperature (°C) and t represents the time (min) of the HTC experiments.

The product yield was calculated using the following equation:

$$\text{product yield } (\%) = \frac{\text{mass of desired product (g)}}{\text{mass of initial wet OPW (g)}} \times 100 \quad (1)$$

The solid hydrochar was sequentially washed with warm distilled water and finally dried overnight under vacuum at 100 °C. The obtained hydrochar samples are hereafter named as S-HC T-t, where T denotes the reaction temperature (°C) and t represents the time (min) of the HTC experiments.

The hydrochar samples, obtained in the experiments performed at different initial pH values at 180 °C and 60 min, were designated as SA-HC1, SA-HC2, AA-HC1, AA-HC2 and AA-HC3, where the number represents the pH value theoretically calculated, and the prefixes SA and AA refer respectively to sulfuric acid and acetic acid used as acidifying agents.

The mass yield of HC (MY) was calculated using the following equation:

$$\text{mass yield, wt \%} = \frac{M_{\text{Hydrochar, g(db)}}}{M_{\text{Feedstock, g(db)}}} \times 100 \tag{2}$$

It is well known that the relationship between temperature and time, defined and quantified by Overend and Chornet using the severity coefficient (R_0) based on the assumptions of first-order kinetics and Arrhenius temperature behavior of the aqueous pre-hydrolysis of Kraft pulping, greatly influences the physicochemical properties of the lignocellulosic substrates during subcritical and supercritical water treatments [26]. Therefore, the role of time and temperature on the HTC of OP was interpreted in terms of (R_0), expressed as

$$\log_{10} R_0 = \log_{10}[t \times \exp(\frac{T - T_0}{14.75})] \tag{3}$$

where t is the time (min), T is the temperature (°C), T_0 is the reference temperature generally set at 100 °C. Assuming the overall reaction following first-order kinetics and Arrhenius relation of temperature, the empirical parameter ω is the fitted parameter, which in this and most other studies is assigned the value of 14.75 and equates to a reaction that doubles in reaction rate for every 10 °C increase in temperature. The ω value in severity function Equation (3) was shown to be inversely proportional to activation energy, as expressed below (4):

$$\omega = \frac{Tf^2 R}{Ea} \tag{4}$$

where Tf is the temperature in the middle of the range of experimental conditions (floor temperature); R is the universal gas constant; Ea is the apparent activation energy.

2.3. Characterization

2.3.1. Characterization of the Bio-Oil Fraction

Products in the aqueous liquid phase as obtained after the separation of the solid carbonaceous fraction were quantified by using an off-line Shimadzu HPLC equipped with an Aminex HPX-87-H column using the following parameters: mobile phase 5 mM H_2SO_4 at a speed flow of 0.6 mL min^{-1} and the oven-heated at 70 °C. Every measurement was performed in 30–60 min [27].

The chemical content of the extracted light bio-oil was analyzed by using a GCMS-QP2010 system (Shimadzu, Japan) equipped with a split–splitless injector. A HP-5 weak polar capillary column, 30 m × 0.25 mm i.d. × 0.32 µm film thickness, was used for each GC analysis. The carrier gas was helium at a constant flow rate of 24.3 mL/min. For the setup of each system, 0.5 µL of the GC sample was injected in split mode, with a split ratio of 1:10 for the sample injection into the column. The oven temperature started at 40 °C (held for 1 min) and then was increased to 300 °C (maintained for 5 min), with a heating speed of 3 °C/min. The mass spectrometer operated in electron ionization mode at 70 eV and mass spectra were obtained in a molecular mass range (m/z) of 40–660. The temperature of the transfer line was 250 °C. Compound identification was performed by comparison with spectra obtained from the US National Institute of Standards and Technology (NIST) mass spectral library (ver. 11).

2.3.2. Characterization of the Hydrochar Fraction

Fourier transform infrared (FTIR) spectra were registered using a Perkin Elmer Spectrum 100 spectrometer, furnished with a common ATR sampling accessory. Spectra were recorded at room temperature from 4000 to 600 cm^{-1} and with a resolution of 4.0 cm^{-1}, without any earlier handling. The morphology of hydrochar samples was investigated using a Zeiss 1540XB FE SEM (Zeiss, Germany) instrument operating at 10 kV [28]. The crystalline structure of synthesized materials was investigated using the X-ray powder diffraction (XRD) by means of a Bruker D8 Advance A 25 X-ray diffractometer operating at 40 kV and in the range 20–80° (2-theta), with an increasing rate of 0.01°/s and the recorded diffractograms were deconvolved via an OriginPro 2018 software. B.E.T. surface area and porosity of samples were evaluated by nitrogen adsorption and desorption isotherms carried out at 77 K by a Quantachrome® ASiQwin™ instrument (Anton Paar Companies, Graz, Austria). The thermal stability in the air of investigated hydrochar samples was evaluated by thermogravimetry (TGA) carried out with TA Instruments SDTQ 600 (balance sensitivity: 0.1 mg). Samples (~15 mg) were heated at 20 °C/min from 100 °C up to 1000 °C using a constant air flow rate (100 mL/min), after a preliminary sample stabilization for 30 min at 100 °C to remove the eventually adsorbed water [29]. The weight loss (%) was calculated.

3. Results and Discussion

3.1. Hydrochar Solid Fraction

3.1.1. Hydrochar Yields and Chemical-Structural Characteristics

In order to analyze separately the effect of reaction temperature and time on hydrochar yields, a series of experiments at fixed solid to water ratio were carried out [30]. After 60 min of reaction, the hydrochar yield rose with the reaction temperature from 18.8% at 180 °C up to a maximum of 30.4% at 300 °C (Figure 1). The reaction time provided less decomposition of biomass than the temperature. It can be assumed that for relatively short reaction times of 60 min and for low reaction temperatures, the conversion of the initial biomass is still at a germinal stage, as evidenced by the SEM analysis reported in Figure 1. On the other hand, after only 60 min of reaction, as the temperature increases the conversion process becomes significant, favoring the increase in hydrochar yield.

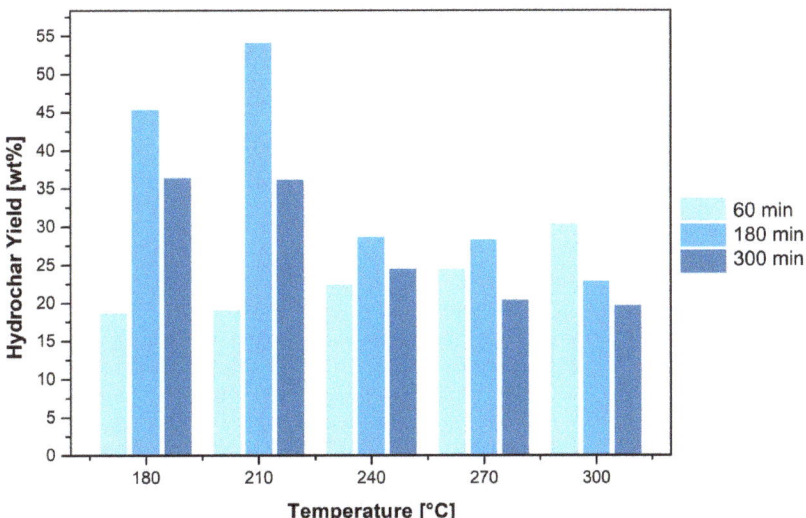

Figure 1. Hydrochar yield as a function of reaction temperature at a fixed reaction time.

After 180 min of reaction, the yield increases with the temperature from 45.4% at 180 °C up to the highest value of 54.1% at 210 °C. Such behavior is in line with the generally accepted reaction mechanism of the HTC process of lignocellulosic materials, where the cellulose degradation starts at a reaction temperature higher than 200 °C while, for a higher reaction temperature, a clear decrease in hydrochar yield is observed as a consequence of gas phase products formation. Accordingly, due to the more relevant volatilization processes under drastic conditions, a notable yield decrease is evident at longer reaction times (300 min) already at 180 °C. Such findings, taking into account data obtained by other authors [30], confirm that process parameters (the reaction time and the reaction temperature) play a fundamental role on the hydrochar yields and cannot be analyzed separately. The reported results differ from those reported by Erdogan et al. [23] where a correlation between the reaction time and the hydrochar yield, in the HTC of the orange pomace in the T range 175–260 °C, was not found. This substantial difference is probably due to the distinctive content of cellulose and hemicellulose and their weight ratio, in the orange peel and in the pomace [31], that probably affects the global carbonization process with a consequent different amount of solid obtained over time, making the results between the two materials difficult to compare.

The simultaneous effect of temperature and reaction time, expressed in terms of $logR_0$, on hydrochar yields is reported in Table 1. A substantial difference was observed in hydrochar yield produced at different $logR_0$. Experimental data, also summarized in Table 1, indicate that a rise in the $logR_0$ exerts a comparable promoting effect on the hydrochar production, resulting in a peculiar increasing trend with the severity coefficient and attaining a plateau of ca. 54.05 wt% at log R_0 equal to 5.49 (210 °C at 180 min). A further increase of the $logR_0$ causes a hydrochar yield decrease, with the lowest value of 19.69% found at the highest severity of 8.36 (300 °C at 300 min).

Table 1. Hydrochar yields as a function of $logR_0$.

Sample	$LogR_0$	HC Yields (wt%)
S-HC$_{150-60}$	3.25	13.27
S-HC$_{180-60}$	4.13	18.74
S-HC$_{180-180}$	4.61	35.37
S-HC$_{180-300}$	4.83	36.40
S-HC$_{210-60}$	5.01	49.10
S-HC$_{210-180}$	5.49	54.05
S-HC$_{210-300}$	5.71	36.13
S-HC$_{240-60}$	5.90	29.45
S-HC$_{240-180}$	6.37	28.66
S-HC$_{240-300}$	6.59	24.48
S-HC$_{270-60}$	6.78	24.53
S-HC$_{270-180}$	7.26	22.34
S-HC$_{270-300}$	7.48	20.39
S-HC$_{300-60}$	7.66	20.40
S-HC$_{300-180}$	8.14	20.87
S-HC$_{300-300}$	8.36	19.69

A similar trend in the simultaneous effect of temperature and reaction time upon energy densification and mass yield have been reported by other researchers using different biomass raw materials [32–34], supporting that higher temperature and reaction time lead to an enrichment in fixed carbon and a consequent higher heat production but lower solids recovery, probably due to a higher conversion of cellulose, hemicellulose and lignin. This allows for a series of deoxygenating processes (e.g., dehydration, decarboxylation), resulting in the formation of organic matter and a decrease in oxygen and hydrogen content [31].

Termogravimetric (TGA) and energy-dispersive spectroscopy (EDS) analysis (shown in Figures S1 and S2 in ESI) of the different hydrochar samples, obtained under different hydrothermal temperatures at a fixed residence time of 180 min, confirm the noticeable role of reaction temperature in the yield and elemental composition of hydrochar. Apart from the S-HC$_{180-180}$ sample, the breakdown of all samples in the air's atmosphere takes place in three stages. Until 200 °C the weight loss (stage I) is very low (less than 2%), probably due to the hydrophobic nature of the hydrochars, which hinders the absorption of a large water quantity inside their structures. When increasing the temperature a large weight loss (stage II) starts, centered at around 300 °C and associated with the release of organic volatile matters due to the decomposition of hemicellulose and cellulose. The successive stage (III) is related to surface and bulk hydrochar oxidation and suggests that the progressive loosing of the organic constituents, originally present in the orange peels, and the surface groups of hydrochar are formed at different hydrothermal temperatures, leading to samples production characterized by a different chemical composition.

Furthermore, EDS spectra of S-HC$_{210-60}$ reveals that at a lower temperature of reaction, many residual oxygenated groups still remain on the surface, evidenced by a high content of oxygen (about 20 wt%) and a lower content of carbon (ca. 78 wt%). On the contrary, an increasing of the reaction temperature up to 300 °C causes a decrease of the oxygen content (12.1%), ascribed to dehydration processes from cellulose to hydrochar [35] and a consequent enhancement of the carbon content (ca. 88 wt%).

As above mentioned, the reaction temperature influences not only the yields of the obtained hydrochar samples but also its morphology and composition, as revealed by SEM experiments. The morphology of samples prepared at lower hydrothermal temperature (Figure 2A) is not very dissimilar to that of the pectin and lignocellulose (the main components of the orange peels raw material) [36]. On the other hand, at 300 °C the material produced shows the presence of the characteristic hydrochar microspheres.

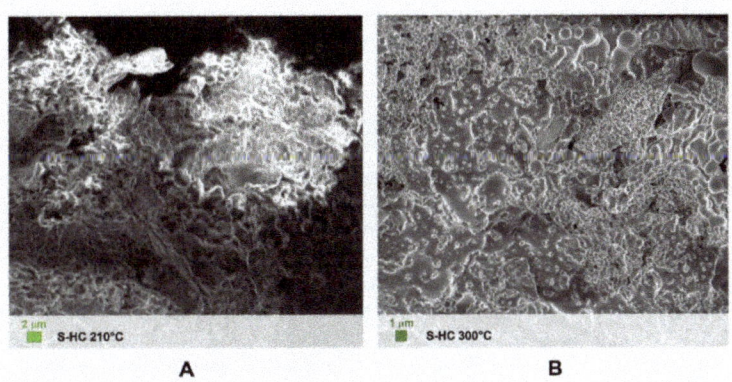

Figure 2. SEM analysis of (**A**) S-HC$_{210-60}$ and (**B**) S-HC$_{300-60}$.

XRD analysis was used for the determination of crystalline structures in the hydrochar samples (Figure 3).

The reflections of crystalline cellulose at 16 and 22.6 theta are noticeably evidenced in hydrochars obtained at the lowest temperature. As the temperature of the hydrochar preparation increases, a gradual shift from orderly crystallites to transition crystallites can be noticed in XRD spectra. Upon further increases of the hydrothermal temperature, these peaks disappear, whereas new broad peaks, at around 25 and 40 2-theta (*), related to turbostratic structure of disordered carbon coming from the (100) plane of graphite appear and gradually grow in intensity. Further increasing of the temperature to 300 °C leads to a complete conversion of the "crystalline stage" to the "amorphous stage", as diffraction peaks at 300 °C suffer from broad shape, low intensities and low signal to noise ratio. As evidenced in Figure S3, the reaction time is irrelevant on the crystalline structure of the samples obtained at different reaction temperatures. This phenomenon can be explained by considering that the crystalline structure of cellulose in the lignocellulosic biomass has well-packed long chains characterized by strong hydrogen bonding networks, which maintain the sugar ring assembly promoted by the hydrolysis reaction occurring during the hydrothermal carbonization process [35].

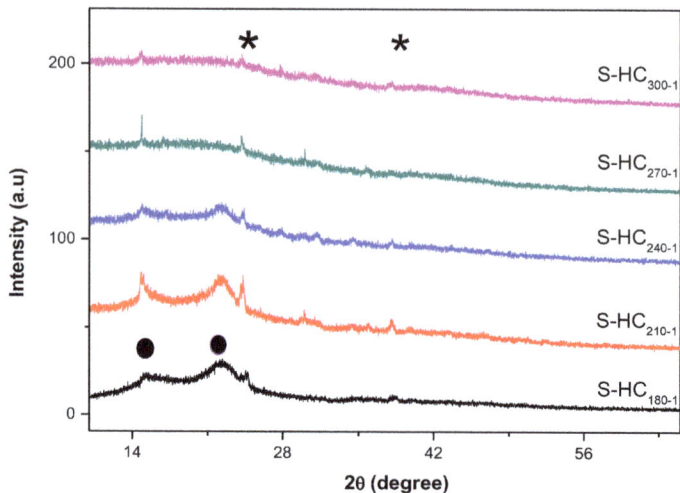

Figure 3. X-ray diffractograms of the hydrochar obtained at different temperatures after 60 min of reaction time. (● reflections of crystalline cellulose; * reflections of turbostratic structure of disordered carbon).

FTIR spectra at different reaction temperatures and reaction times are reported in Figure S4 in ESI. As can be seen from the FT-IR spectra, the reaction time does not affect the structural properties of the samples obtained at different reaction temperatures, therefore an example of a summary of the structural properties and N_2 adsorption-desorption isotherms obtained at various temperatures and at a fixed reaction time of 60 min is reported in Table 2. The adsorption bands at 1608 cm^{-1} and 1701 cm^{-1} correspond to the C=C vibrations and the C=O band, respectively, displaying the asymmetric stretch of aromatic rings, carbonyl, quinone, ester or carboxylic groups probably with a small quantity of amide, revealing the decarboxylation reaction and the aromatization of the orange peel waste during the hydrothermal process. The peaks at 1026 cm^{-1} can be attributed to the C–O stretching vibrations of carboxylic acids and esters or carbohydrates, while the band at about 3300 cm^{-1} suggests the O–H stretching of hydroxylic or caboxylic functionalities. Aliphatic C–H bands are found at 3000–2800 cm^{-1}. The spectrum of S-HC$_{180-60}$ reveals a peak at 1120–1050 cm^{-1}, possibly related to a C–O band of the lignocellulose component of the orange peels still present. According to XRD data, this peak decreases in intensity when increasing the treatment temperature. The presence of a peak at 1527 cm^{-1} strongly suggests the presence of a nitro-compound characterized by the N–O asymmetric stretching. The peak at 1431 cm^{-1} may also correspond to the asymmetric and symmetric stretching of the carboxylate (COO–) group. The fragment values became stronger and broader when experiments are carried out at a higher temperature, in agreement with TGA and EDS analysis confirming that during the HTC process a series of dehydration, decarboxylation and aromatization reactions occur, with a consequent decrease in the hydrochar yield as well as an enrichment in C content.

Moreover, the presence of an absorptions peak at 1026 cm^{-1} in the S-HC$_{180-60}$ sample, related to the C–O stretching vibration, can be again related to the presence of small amount of crystalline cellulose. As expected, upon increasing the reaction temperature, the absence of such a peak is indicative of the full conversion of OPW into biochar and bio-oil. Values of BET surface area and pore volume, reported in Table 2, relative to S-HC samples, at various temperatures and at a fixed reaction time of 60 min in accordance with other hydrochar prepared in similar experimental conditions [37] show that both parameters increase about three–four times when increasing the reaction temperature going from 180 to 300 °C.

Table 2. FTIR assignments and structural properties of the hydrochar samples at different reaction temperatures and a fixed reaction time of 60 min.

Observed Peak Intensity	Possible Functional Groups	Sample	B.E.T. S.A. (m^2/g)	Pore Volume (cc/g)	Pore Radius Dv(r) (Å)
3300 cm^{-1}	O–H (alcohols, phenols, carboxylic acid)	S-HC$_{180-60}$	4.9	0.009	17.9
3000 cm^{-1}	C–H (aliphatic methyl)	S-HC$_{210-60}$	5.5	0.008	20.4
1701 cm^{-1}	C=O (ketone, aldehydes, amides)	S-HC$_{240-60}$	7.7	0.010	17.6
1608 cm^{-1}	C=C (aromatic rings, carbonyl, quinone, ester or carboxyl groups)	S-HC$_{270-60}$	9.1	0.011	17.8
1527 cm^{-1}	N–O (nitro)				
1431 cm^{-1}	COO– (carboxylate)	S-HC$_{300-60}$	18.4	0.029	22.4
1120 cm^{-1}	C–O of lignocellulose				
1026 cm^{-1}	C–O (carboxylic acid, esters)				

3.1.2. Effect of Initial pH and Solid to Water Ratio

Although the more relevant literature results identify reaction temperature and residence time as the main process parameters affecting the amount and chemical composition of the hydrochar produced, it is worth mentioning that both the solid to water ratio and the initial pH could play an important role, given that one of the goals of the HTC process is to break down the unbending structure of the starting biomass material into small and lower molecular weight chains. In fact, it is well known that addition of sulfuric acid or acetic acid in the reaction mixture can positively influence the HTC process by catalyzing hydrolysis reactions of cellulose and hemicellulose formed in the experimental conditions usually adopted. Regarding the effect of the initial pH, results reported in Table 3 show that the initial pH impacts the hydrochar yield, increasing as the higher acid concentration increases, with the highest yield of 30.12 wt% attained for SA-HC1, almost 50% higher than that observed in the reference experiment (180 °C, 60 min). However, these change slightly as the acetic acid concentration increases. Incidentally, we observe that CH_3COOH is a weak acid whereas H_2SO_4 is a strong acid when the first dissociation constant is concerned. Results also suggest that rather than the pH itself, it is the type of additive used that affects the yield. Indeed, the change in hydrochar yields could be explained considering that during the sulfuric acid catalyzed hydrolysis, insoluble humins indistinguishable from HC products are formed, with a subsequent enhancement of the number of solid products obtained.

Table 3. Hydrochar yields as a function of initial pH.

Sample	pH [1]	pH [2]	Hydrochar Yield (wt%)
S-H$_{180-60}$	-	3.60	18.74
SA-HC1	1	1.61	30.62
SA-HC2	2	1.89	27.36
AA-HC1	1	1.20	21.51
AA-HC2	2	1.80	19.39
AA-HC3	3	2.40	15.24

[1] theoretically calculated; [2] experimentally measured.

On the contrary, the acetic acid is a natural by-product of HTC of lignocellulosic feedstocks, due to the hydrolysis and dehydration of cellulose and hemicellulose in the presence of subcritical water into short-chain organic acids: primarily acetic, formic, and lactic acid [38,39]. Therefore, a further addition of acetic acid in the reaction media, favoring the degradation of biomass components, leads to a higher heat production and a lower hydrochar yield.

Results obtained in this study (see Table S1 in ESI) also indicate that the role of the biomass/water ratio on hydrochar yield is insignificant for all values investigated. This finding is in agreement with the results reported by A. Toptas Tag et al. [40] for the HTC of sunflower stalk as agricultural waste, poultry litter as animal waste, and algal biomass by Sapio et al. [41], but disagree with data reported by other authors [42,43]. In any case, in the literature there is no unanimous opinion on the role played by the different water/biomass ratio, and generally a fixed biomass to water ratio has been used by many authors. However, all results obtained as a function of the reaction temperature, reaction time, biomass/water ration and pH confirm that it is not always easy to make a direct comparison between the hydrochar yields reported in the literature and those reported in this study since there are many parameters that can affect the process, such as the experimental setup, the type of feedstock, the total solid and water amount, and the final amount of the hydrochar produced.

3.2. Hydrothermal Bio-Oil Liquid Fraction

3.2.1. Composition of Hydrothermal Bio-Oil Liquid Fraction

The general chemical composition of the light bio-oil obtained at different reaction temperatures and at a fixed residence time of 60 min was analyzed by GC-MS (Table 4).

Table 4. Major chemical components of the light bio-oil determined by GC-MS analysis.

		Sample			
	Compounds Name	L-HC$_{150-60}$	L-HC$_{180-60}$	L-HC$_{240-60}$	L-HC$_{300-60}$
		Peak Area%			
Furans	Furfural	18.94	16.69	1.63	-
	2-Furancarboxaldehyde, 5-methyl	14.86	7.88	5.05	-
	5-Hydroxymethylfurfural	41.97	61.82	28.61	-
Phenols	Phenol	0.57	0.10	4.72	9.62
	Catechol	0.60	0.29	4.52	8.64
	1,2-Benzenediol, 3-methyl	-	0.11	0.78	1.04
	Hydroquinone	-	-	1.91	7.26
	p-Cresol	-	-	0.79	2.17
	Phenol, 2-methyl	-	-	0.27	0.69
Acids	Benzoic acid	2.27	0.61	2.67	6.21
	2-Pentenoic acid	-	-	0.32	-
Ketones	2-Pentanone, 4-hydroxy-4-methyl	0.40	0.68	3.87	1.20
	Ethanone, 1-(2-furanyl)	0.69	0.46	0.40	1.86
	1,2-Cyclopentanedione, 3-methyl	0.63	0.81	-	-
	2-Cyclopenten-1-one, 2-hydroxy-3-methyl	-	-	4.37	0.99
	2-Cyclopenten-1-one, 2-methyl	-	-	0.61	5.83
	2-Cyclopenten-1-one, 3-methyl	-	-	0.61	7.41
Aldehydes	1H-Pyrrole-2-carboxaldehyde	0.86	0.41	0.66	-
	Benzaldehyde, 3-hydroxy	-	0.17	0.83	-
	Vanillin, acetate	-	-	1.07	-
Alcohols	α-Terpineol	0.91	0.24	-	1.01
	Benzyl alcohol	-	-	0.29	-
	3-Pyridinol	-	-	10.14	4.63
Alkenes	1-Nonadecene	0.24	0.33	1.65	-
	1-Pentadecene	-	0.59	3.51	-
	1-Heptadecene	-	-	3.64	-

The reaction mechanism involved in the formation of bio-oil during the HTC of various lignocellulosic biomasses has been described by several authors [38,44] and consists of a series of consecutive and parallel reactions starting with acid hydrolysis of polysaccharides to form monosaccharides. Glucose and fructose can be dehydrated by acids into 5-hydroxymethylfurfural and furfural. Furans and other reaction intermediates can undergo further transformation processes (e.g., isomerization, condensation, rehydration, hydrations) or they can be degraded into humins [45].

As illustrated in Table 4, the light bio-oil is mainly divided into seven categories: furans, phenols, acids, ketones, aldehydes, alcohols and alkenes (Figure S5 in ESI). At lower HT temperatures, the main compounds at 200 °C are furan derivatives. When increasing the temperature, the composition of the light bio-oil becomes more complex, showing a higher concentration of acid phenolic compounds as a consequence of humins formation.

3.2.2. Production of Furan Derivatives and Levulinates from Hydrothermal Upgrading of Orange Peel Waste

To investigate the best reaction conditions that maximize the production of furan and levulinate derivatives, a systematic study on the effect of (i) time; (ii) temperature; (iii) initial acid concentration; and (iv) co-solvent was carried out.

Reaction temperature is surely a crucial parameter for investigating the overall productivity of furan derivatives starting from OPW. The yield profiles of FU, 5-HMF and levulinic acid are reported in Figure 4. Under neutral conditions, FU and 5-HMF yields gradually increase, reaching the highest value at 180 °C. The observed decrease in the production of furans at higher reaction temperatures is in line with other reports [46] and can be related to the formation of humin type by-products. Indeed, dark-brown insoluble products were formed in experiments carried out at 240 °C and 300 °C. As expected, under the reaction conditions adopted very low concentrations of levulinic acid (LA) were registered in all investigated reaction temperatures. A noticeable amount of LA can be seen at 300 °C due to the higher concentration of protons driving from the dissociation promoted by high reaction temperatures.

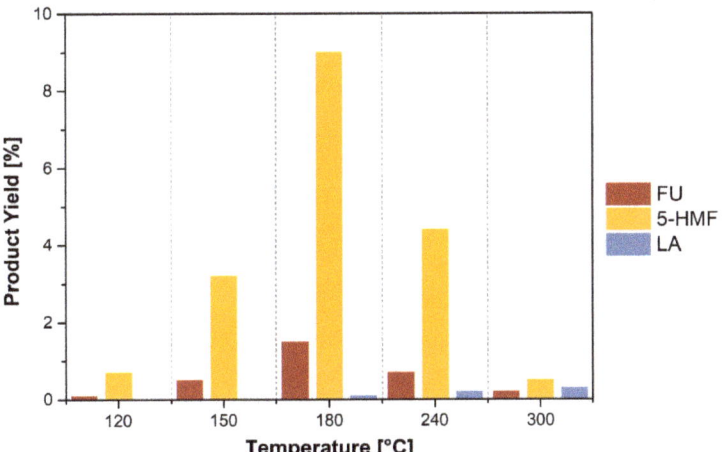

Figure 4. Effect of HTC reaction temperature on the production of furfural (FU), 5-hydroxymethylfurfural (5-HMF) and levulinic acid (LA) from OPW under neutral conditions (Reaction conditions: 20 g of OPW; 20 mL of H_2O; time: 60 min; autogenous pressure; stirring: 600 rpm).

Having found 180 °C as the best reaction temperature that maximizes FU derivatives, the conversion of OPW was also investigated at different reaction times (15, 30, 60, 180 and 300 min) (Figure 5). The highest 5-HMF and FU yields were already gained after only 30 min. A slight decrease in the production of furans was registered after 360 min and may be related both to the thermal degradation of furan derivatives as well as to the fact that a prolonged hydrolysis time increases formation of humins [47].

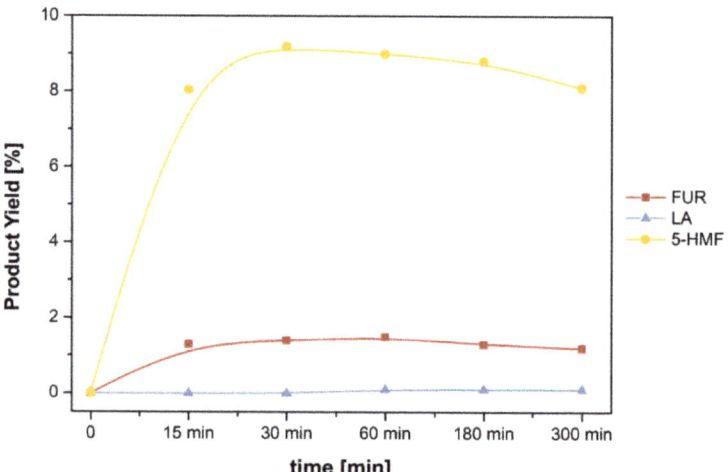

Figure 5. Effect of HTC reaction time on the production of furfural (FU), 5-hydroxymethylfurfural (5-HMF) and levulinic from (LA) OPW under neutral conditions (Reaction conditions: 20 g of OPW; 20 mL of H_2O; temperature: 180 °C; autogenous pressure; stirring: 600 rpm).

On the other hand, by changing the conditions using a sulfuric acid solution as reaction media, a decrease in the production of furans together with a higher production of levulinic acid is observed as a consequence of the acid hydrolysis (Figure 6). Upon increasing the H_2SO_4 content (from 0.2 to 1.0 M), a slight decrease in the overall LA can be noticed.

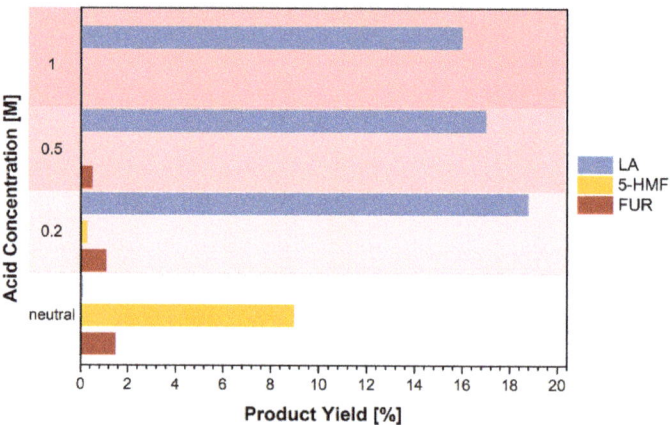

Figure 6. Effect of acid conditions on the production of furfural (FU), 5-hydroxymethylfurfural (5-HMF) and levulinic acid (LA) (Reaction conditions: 20 g of OPW; 20 mL solution of H_2SO_4 (from 0.1 to 1.0 M) in H_2O; temperature: 180 °C; time: 60 min; autogenous pressure; stirring: 600 rpm).

Analogous results are registered in the presence of simple alcohols as co-solvents, such as methanol and ethanol, that permit the direct formation of alkyl levulinates (Figure 7), used as flavoring/fragrance agents or as fuel bio-additives [48,49]. Moreover, methyl- and ethyl-levulinate now represent a valid starting bio-based feedstock for the preparation of γ-valerolactone [50,51] and find many applications ranging from flavoring agent to green solvent or as an intermediate in the synthesis of bio-based chemicals and polymers [52,53].

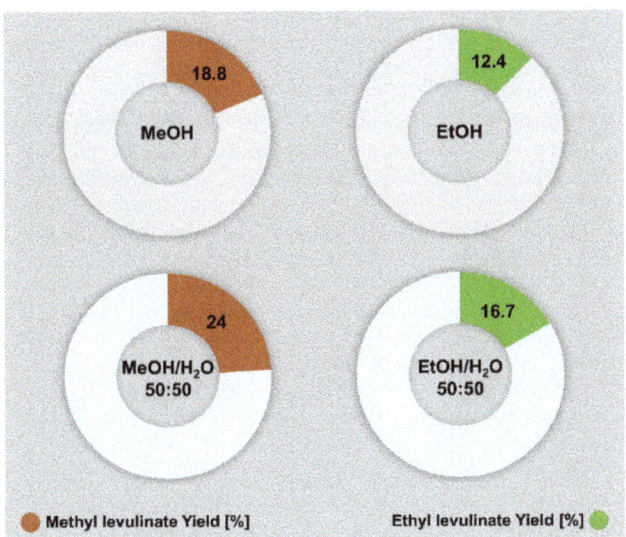

Figure 7. Co-solvent effect of HTC reaction temperature on the production of alkyl levulinates (Reaction conditions: 20 g of OPW; temperature: 180 °C; time: 60 min; autogenous pressure; stirring: 600 rpm).

Indeed, the best results in terms of alkyl-levulinate production were obtained when using a 0.2 M sulphuric acid solution with higher yields from the methanol esterification that, being characterized by a shorter carbon chain, is definitely a better entering group than ethanol (Table 5).

Table 5. Effect of acid conditions on the preparation of methyl-levulinate and ethyl-levulinate starting from OPW by using HTC process (Reaction conditions: 20 g of OPW; temperature: 180 °C; time: 60 min; autogenous pressure; stirring: 600 rpm).

Product	Solvent (wt:wt)	(0.1 M H_2SO_4)	(0.1 M H_2SO_4)	(0.1 M H_2SO_4)
Methyl-levulinate	water: methanol (50:50)	7.1 (% yield)	24.0 (% yield)	14.2 (% yield)
Ethyl-levulinate	water: ethanol (50:50)	4.9 (% yield)	16.7 (% yield)	7.5 (% yield)

4. Conclusions

In this work, we demonstrated that the hydrothermal carbonization process can be successfully adopted for the complete upgrading of the orange peel waste into hydrochar and value-added chemicals. The main processing variables, reaction temperature, initial pH and residence time affect the mass yield (MY), while the solid to liquid ratio was found to be insignificant for all L/S investigated. Indeed, there was a strong correlation between temperature and residence time, suggesting that the role of these two variables cannot be analyzed independently. The highest yield of hydrochar was obtained at a 210 °C reaction temperature, 180 min residence time, 6/1 w/w orange peel waste to water ratio and a 3.6 initial pH. The results suggest that the conversion of the citrus waste occurs during the hydrothermal carbonization process due to a series of reactions such as decarboxylation and dehydration, leading to an improvement of the chemical, structural and morphological characteristics of the optimized hydrochar. This makes it a carbonaceous material suitable for a wide range of applications, both in the chemical and energy fields.

The bio-products distribution strongly depends on the applied reaction conditions. Overall, 180 °C was found to be the best reaction temperature that maximizes the production of furfural and 5-HMF in the presence of pure water as a reaction medium. On the other hand, by using a sulfuric acid solution as reaction media, levulinic acid can be easily obtained as a main product. Accordingly, a good production in methyl levulinate and ethyl levulinate can be achieved in the presence of methanol or ethanol, respectively, with the best yield obtained with methanol in the esterification reaction. Future research will be devoted to a prior recovery of more valuable compounds from OPW (e.g., limonene, pectin) before their hydrothermal conversion into hydrochar and bulk chemicals with a cascade high-to-low value approach.

Supplementary Materials: The following are available online at https://www.mdpi.com/article/10.3390/app112210983/s1. Figure S1. TGA curve in air of the hydrochar samples. Figure S2. SEM-EDS analyses of (a) S-HC$_{210-60}$ and (b) S-HC$_{300-60}$. The inset shows the elemental analysis of these samples. Figure S3. X-ray diffrattograms of the hydrochar obtained at different temperature after (a) 180 min of reaction time and (b) 300 min of reaction time. Figure S4. FT-IR analysis of the hydrochar samples at (a) 60 min of reaction time; (b) 180 min of reaction time; (c) 300 min of reaction time. Table S1. Hydrochar yield as function of L/S ratio.

Author Contributions: Conceptualization: E.P. and C.E.; methodology: E.P., P.S.C., D.I., F.M. (Francesco Mauriello) and C.E.; formal analysis: A.S., V.B. and F.M. (Federica Marra); writing—original draft preparation: all authors; writing—review and editing. All authors have read and agreed to the published version of the manuscript.

Funding: This research received no external funding.

Institutional Review Board Statement: Not applicable.

Informed Consent Statement: Not applicable.

Data Availability Statement: The data presented in this study are available on request from the corresponding author.

Conflicts of Interest: The authors declare no conflict of interest.

References

1. FAO. *Fruit and Vegetables-Your Dietary Essentials. The International Year of Fruits and Vegetables*; FAO: Rome, Italy, 2021.
2. Choi, I.S.; Lee, Y.G.; Khanal, S.K.; Park, B.J.; Bae, H.J. A low-energy, cost-effective approach to fruit and citrus peel waste processing for bioethanol production. *Appl. Energy* **2015**, *140*, 65–74. [CrossRef]
3. FAS-USDA (Foreign Agricultural Service-United States Department of Agriculture). Citrus: World Markets and Trade. 2014. Available online: https://apps.fas.usda.gov/psdonline/circulars/citrus.pdf (accessed on 20 July 2020).
4. Satari, B.; Karimi, K. Citrus processing wastes: Environmental impacts, recent advances, and future perspectives in total valorization. *Resour. Conserv. Recycl.* **2018**, *129*, 153–167. [CrossRef]
5. Zema, D.A.; Calabrò, P.S.; Folino, A.; Tamburino, V.; Zappia, G.; Zimbone, S.M. Valorisation of citrus processing waste: A review. *Waste Manag.* **2018**, *80*, 252–273. [CrossRef]
6. Raimondo, M.; Caracciolo, F.; Cembalo, L.; Chinnici, G.; Pecorino, B.; D'Amico, M. Making virtue out of necessity: Managing the citrus waste supply chain for bioeconomy applications. *Sustainability* **2018**, *10*, 4821. [CrossRef]
7. De la Torre, I.; Martin-Dominguez, V.; Acedos, M.G.; Esteban, J.; Santos, V.E.; Ladero, M. Utilisation/upgrading of orange peel waste from a biological biorefinery perspective. *Appl. Microbiol. Biotechol.* **2019**, *103*, 5975–5991. [CrossRef] [PubMed]
8. Fazzino, F.; Mauriello, F.; Paone, E.; Sidari, R.; Calabrò, P.S. Integral valorization of orange peel waste through optimized ensiling: Lactic acid and bioethanol production. *Chemosphere* **2021**, *271*, 129602. [CrossRef]
9. Fidalgo, A.; Ciriminna, R.; Carnaroglio, D.; Tamburino, A.; Cravotto, G.; Grillo, G.; Ilharco, L.M.; Pagliaro, M. Eco-friendly extraction of pectin and essential oils from orange and lemon peels. *ACS Sustain. Chem. Eng.* **2016**, *4*, 2243–2251. [CrossRef]
10. Espro, C.; Paone, E.; Mauriello, F.; Gotti, R.; Uliassi, E.; Bolognesi, M.L.; Rodriguez Padron, D.; Luque, R. Sustainable production of pharmaceutical, nutraceutical and bioactive compounds from biomass and waste. *Chem. Soc. Rev.* **2021**, *50*, 11191. [CrossRef] [PubMed]
11. Burguete, P.; Corma, A.; Hitzl, M.; Modrego, R.; Ponceb, E.; Renz, M. Fuel and chemicals from wet lignocellulosic biomass waste streams by hydrothermal carbonization. *Green Chem.* **2016**, *18*, 1051–1060. [CrossRef]
12. Fang, J.; Zhan, L.; Ok, Y.S.; Gao, B. Mini review of potential applications of hydrochar derived from hydrothermal carbonization of biomass. *J. Ind. Eng. Chem.* **2018**, *57*, 15–21. [CrossRef]

13. Satira, A.; Espro, C.; Paone, E.; Calabrò, P.S.; Pagliaro, M.; Ciriminna, R.; Mauriello, F. The limonene biorefinery: From extractive technologies to its catalytic upgrading into p-cymene. *Catalysts* **2021**, *11*, 387. [CrossRef]
14. Xiao, K.; Liu, H.; Li, Y.; Yang, G.; Wang, Y.; Yao, H. Excellent performance of porous carbon from urea-assisted hydrochar of orange peel for toluene and iodine adsorption. *Chem. Eng. J.* **2020**, *382*, 122997. [CrossRef]
15. Kalderis, D.; Papameletiou, G.; Kayan, B. Assessment of orange peel hydrochar as a soil amendment: Impact on clay soil physical properties and potential phytotoxicity. *Waste Biomass* **2019**, *10*, 3471–3484. [CrossRef]
16. Pistone, A.; Espro, C. Current trends on turning biomass wastes into carbon materials for electrochemical sensing and rechargeable battery applications. *Curr. Opin. Green Sustain. Chem.* **2020**, *26*, 100374–100381. [CrossRef]
17. Gou, H.; He, J.; Zhao, G.; Zhang, L.; Yang, C.; Rao, H. Porous nitrogen-doped carbon networks derived from orange peel for high-performance supercapacitors. *Ionics* **2019**, *25*, 4371–4380. [CrossRef]
18. Bressi, V.; Ferlazzo, A.; Iannazzo, D.; Espro, C. Graphene quantum dots by eco-friendly green synthesis for electrochemical sensing: Recent advances and future perspectives. *Nanomaterials* **2021**, *11*, 1120. [CrossRef]
19. Espro, C.; Satira, M.; Anajafi, M.K.; Iannazzo, D.; Neri, G. Orange peels-derived hydrochar for chemical sensing applications. *Sens. Actuators B Chem.* **2021**, *341*, 130016–130027. [CrossRef]
20. Lathiya, D.R.; Bhatt, D.V.; Maheria, K.C. Synthesis of sulfonated carbon catalyst from waste orange peel for cost effective biodiesel production. *Bioresour. Technol. Rep.* **2018**, *2*, 69–76. [CrossRef]
21. Singh, M.; Pandey, N.; Dwivedi, P.; Kumar, V.; Mishra, B.B. Production of xylose, levulinic acid, and lignin from spent aromatic biomass with a recyclable Brønsted acid synthesized from d-limonene as renewable feedstock from citrus waste. *Bioresour. Technol.* **2019**, *293*, 122105. [CrossRef]
22. Pileidis, F.D.; Tabassum, M.; Coutts, S.; Titiric, M.-M. Esterification of levulinic acid into ethyl levulinate catalysed by sulfonated hydrothermal carbons. *Chin. J. Catal.* **2014**, *35*, 929–936. [CrossRef]
23. Erdogan, E.; Atila, B.; Mumme, J.; Reza, M.T.; Toptas, A.; Elibol, M.; Yanik, J. Characterization of products from hydrothermal carbonization of orange pomace including anaerobic digestibility of process liquor. *Bioresour. Technol.* **2015**, *196*, 35–42. [CrossRef]
24. Aboagye, D.; Banadda, N.; Kiggundu, N.; Kabenge, I. Assessment of orange peel waste availability in Ghana and potential bio-oil yield using fast pyrolysis. *Renew. Sustain. Energy Rev.* **2017**, *70*, 814–821. [CrossRef]
25. Xu, C.; Paone, E.; Rodríguez-Padrón, D.; Luque, R.; Mauriello, F. Recent catalytic routes for the preparation and the upgrading of biomass derived furfural and 5-hydroxymethylfurfural. *Chem. Soc. Rev.* **2020**, *49*, 4273–4306. [CrossRef]
26. Overend, R.P.; Chornet, E. Fractionation of lignocellulosics by steam-aqueous pretreatments. *Philos. Trans. R. Soc. Lond. Ser. A Math. Phys. Sci.* **1987**, *321*, 523–536.
27. Gumina, B.; Espro, C.; Galvagno, S.; Pietropaolo, R.; Mauriello, F. Bioethanol production from unpretreated cellulose under neutral selfsustainable hydrolysis/hydrogenolysis conditions promoted by the heterogeneous Pd/Fe$_3$O$_4$ catalyst. *ACS Omega* **2019**, *4*, 352–357. [CrossRef] [PubMed]
28. Malara, A.; Paone, E.; Bonaccorsi, L.; Mauriello, F.; Macario, A.; Frontera, P. Pd/Fe$_3$O$_4$ nanofibers for the catalytic conversion of lignin-derived benzyl phenyl ether under transfer hydrogenolysis conditions. *Catalysts* **2020**, *10*, 20. [CrossRef]
29. Malara, A.; Paone, E.; Frontera, P.; Bonaccorsi, L.; Panzera, G.; Mauriello, F. Sustainable exploitation of coffee silverskin in water remediation. *Sustainability* **2018**, *10*, 3547. [CrossRef]
30. Zhuang, X.; Zhan, H.; Songa, Y.; He, C.; Huang, Y.; Yin, X.; Wu, C. Insights into the evolution of chemical structures in lignocellulose and non-lignocellulose biowastes during hydrothermal carbonization (HTC). *Fuel* **2019**, *236*, 960–974. [CrossRef]
31. Rivas-Cantu, R.C.; Jones, K.D.; Mills, P.L. A citrus waste-based biorefinery as a source of renewable energy: Technical advances and analysis of engineering challenges. *Waste Manag. Res.* **2013**, *31*, 413–420. [CrossRef] [PubMed]
32. Resa, M.T.; Yang, X.; Coronella, C.J.; Lin, H.; Hathwaik, U.; Shintani, D.; Neupane, B.P.; Miller, G.C. Hydrothermal carbonization (HTC) and pelletization of two arid land plants bagasse for energy densification. *ACS Sustain. Chem. Eng.* **2016**, *4*, 1106–1114.
33. Yan, W.; Perez, S.; Sheng, K. Upgrading fuel quality of moso bamboo via low temperature thermochemical treatments: Dry torrefaction and hydrothermal carbonization. *Fuel* **2017**, *196*, 473–480. [CrossRef]
34. Yao, Z.; Ma, X.; Lin, Y. Effects of hydrothermal treatment temperature and residence time on characteristics and combustion behaviors of green waste. *Appl. Therm. Eng.* **2016**, *104*, 678–686. [CrossRef]
35. Wang, S.; Dai, G.; Yang, H.; Luo, Z. Lignocellulosic biomass pyrolysis mechanism: A state-of-the-art review. *Prog. Energy Combust. Sci.* **2017**, *62*, 33–86. [CrossRef]
36. Cai, J.; Li, B.; Wang, J.; Zhao, M.; Zhang, K. Hydrothermal carbonization of tobacco stalk for fuel application. *Bioresour. Technol.* **2016**, *220*, 305–311. [CrossRef]
37. Sevilla, M.; Fuertes, A.B. Chemical and structural properties of carbonaceous products obtained by hydrothermal carbonization of saccharides. *Chem. Eur. J.* **2009**, *15*, 4195–4203. [CrossRef]
38. Reza, M.T.; Uddin, M.H.; Lynam, J.G.; Hoekman, S.K.; Coronella, C.J. Hydrothermal carbonization of loblolly pine: Reaction chemistry and water balance. *Biomass Convers. Biorefinery* **2014**, *4*, 311–321. [CrossRef]
39. Reza, M.T.; Rottler, E.; Herklotz, L.; Wirth, B. Hydrothermal carbonization (HTC) of wheat straw: Influence of feedwater pH prepared by acetic acid and potassium hydroxide. *Bioresour. Technol.* **2015**, *182*, 336–344. [CrossRef]
40. Tag, A.T.; Duman, G.; Yanik, J. Influences of feedstock type and process variables on hydrochar properties. *Bioresour. Technol.* **2018**, *250*, 337–344.

41. Sabio, E.; Álvarez-Murillo, A.; Román, S.; Ledesma, B. Conversion of tomato-peel waste into solid fuel by hydrothermal carbonization: Influence of the processing variables. *Waste Manag.* **2016**, *47*, 122–132. [CrossRef] [PubMed]
42. Volpe, M.; Fiori, L. From olive waste to solid biofuel through hydrothermal carbonization: The role of temperature and solid load on secondary char formation and hydrochar energy properties. *J. Anal. Appl. Pyrolysis* **2017**, *124*, 63–72. [CrossRef]
43. Borrero-López, A.M.; Fierro, V.; Jeder, A.; Ouederni, A.; Masson, E.; Celzard, A. High added-value products from the hydrothermal carbonization of olive stones. *Environ. Sci. Pollut. Res.* **2017**, *24*, 9859–9869. [CrossRef] [PubMed]
44. Wu, K.; Gao, Y.; Zhu, G.; Zhu, J.; Yuan, Q.; Chen, J.; Cai, M.; Feng, L. Characterization of dairy manure hydrochar and aqueous phase products generated by hydrothermal carbonization at different temperatures. *J. Anal. Appl. Pyrolysis* **2017**, *127*, 335–342. [CrossRef]
45. Chheda, J.N.; Román-Leshkov, Y.; Dumesic, J.A. Production of 5-hydroxymethylfurfural and furfural by dehydration of biomass-derived mono- and poly-saccharides. *Green Chem.* **2007**, *9*, 342–350. [CrossRef]
46. Puccini, M.; Licursi, D.; Stefanelli, E.; Vitolo, S.; Raspolli Galletti, A.M.; Heeres, H.J. Hydrothermal treatment of orange peel waste for the integrated production of furfural, levulinic acid and reactive hydrochar: Towards the application of the biorefinery concept. *Chem. Eng. Trans.* **2016**, *50*, 223–228.
47. Silva, J.F.L.; Pinto Mariano, A.; Filho, R.M. Less severe reaction conditions to produce levulinic acid with reduced humins formation at the expense of lower biomass conversion: Is it economically feasible? *Fuel Commun.* **2021**, *9*, 100029. [CrossRef]
48. Murat Sen, S.; Gürbüz, E.I.; Wettstein, S.G.; Martin Alonso, D.; Dumesic, J.A.; Maravelias, C.T. Production of butene oligomers as transportation fuels using butene for esterification of levulinic acid from lignocellulosic biomass: Process synthesis and technoeconomic evaluation. *Green Chem.* **2012**, *14*, 3289–3294.
49. Bond, J.Q.; Martin Alonso, D.; Wang, D.; West, R.M.; Dumesic, J.A. Integrated catalytic conversion of gamma-valerolactone to liquid alkenes for transportation fuels. *Science* **2010**, *327*, 1110–1114. [CrossRef]
50. Tabanelli, T.; Paone, E.; Blair Vásquez, P.; Pietropaolo, R.; Cavani, F.; Mauriello, F. Transfer hydrogenation of methyl and ethyl levulinate promoted by a ZrO_2 catalyst: Comparison of batch vs continuous gas-flow conditions. *ACS Sustain. Chem. Eng.* **2019**, *7*, 9937–9947. [CrossRef]
51. Vaśquez, P.B.; Tabanelli, T.; Monti, E.; Albonetti, S.; Dimitratos, N.; Cavani, F. Gas-phase catalytic transfer hydrogenation of alkyl levulinates with ethanol over ZrO_2. *ACS Sustain. Chem. Eng.* **2019**, *7*, 8317. [CrossRef]
52. Wright, W.R.H.; Palkovits, R. Development of heterogeneous catalysts for the conversion of levulinic acid to γ-valerolactone. *ChemSusChem* **2012**, *5*, 1657–1667. [CrossRef]
53. Omoruyi, U.; Page, S.; Hallett, J.; Miller, P.W. Homogeneous catalyzed reactions of levulinic acid: To γ-valerolactone and beyond. *ChemSusChem* **2016**, *9*, 2037–2047. [CrossRef]

Article
Effect of Working Atmospheres on the Detection of Diacetyl by Resistive SnO₂ Sensor

Andrea Gnisci [1,*], Antonio Fotia [2], Lucio Bonaccorsi [3] and Andrea Donato [3]

1. Department of Heritage, Architecture, Urbanism (PAU), University Mediterranea of Reggio Calabria, Via dell'Università 25, 89124 Reggio Calabria, Italy
2. Department of Information Engineering, Infrastructures and Sustainable Energy, University Mediterranea of Reggio Calabria, Via Graziella Loc. Feo di Vito, 89124 Reggio Calabria, Italy; antonio.fotia@unirc.it
3. Department of Civil, Energy, Environment and Material Engineering, University Mediterranea of Reggio Calabria, Via Graziella Loc. Feo di Vito, 89124 Reggio Calabria, Italy; lucio.bonaccorsi@unirc.it (L.B.); andrea.donato@unirc.it (A.D.)
* Correspondence: andrea.gnisci@unirc.it

Abstract: Nanostructured metal oxide semiconductors (MOS) are considered proper candidates to develop low cost and real-time resistive sensors able to detect volatile organic compounds (VOCs), e.g., diacetyl. Small quantities of diacetyl are generally produced during the fermentation and storage of many foods and beverages, conferring a typically butter-like aroma. Since high diacetyl concentrations are undesired, its monitoring is fundamental to identify and characterize the quality of products. In this work, a tin oxide sensor (SnO_2) is used to detect gaseous diacetyl. The effect of different working atmospheres (air, N_2 and CO_2), as well as the contemporary presence of ethanol vapors, used to reproduce the typical alcoholic fermentation environment, are evaluated. SnO_2 sensor is able to detect diacetyl in all the analyzed conditions, even when an anaerobic environment is considered, showing a detection limit lower than 0.01 mg/L and response/recovery times constantly less than 50 s.

Keywords: nanomaterials; MOS; resistive sensor; tin oxide; fermentation; diacetyl

1. Introduction

Gas sensors are of great interest due to their numerous applications and the possibility for a real-time analysis of several analytes [1–3]. Among these, resistive gas sensors exhibit attractive advantages compared with other gas sensors, such as fast and accurate gas detection, flexibility, low cost and small size [4]. The development of high-performance resistive gas sensors requires suitable gas-sensing materials in terms of both physical and chemical properties, such as morphology, crystalline structure, specific surface area and active sites. Since, as is well known, these characteristics all affect the performance of gas sensors, exploring and developing innovative materials has received attention in scientific research in recent years [5]. Attention is concentrated on the development of nanostructured materials, endowed with better sensing properties if compared to the same bulk material, such as carbon-based materials (in the form of carbon dots, nanotubes and graphene) [6–9], polymeric fibers (hybrid nanofibers) [10,11] and metal oxides semiconductor (MOS, as nanospheres, nanowires and nanosheets) [12–17]. Indeed, the large surface areas of nanomaterials ensure more active sites, enabling fast charge transfers and efficient gas-sensitive reactions [4].

Due to their widespread use in a variety of applications, in the recent years time-consuming, complex, and expensive traditional techniques have been substituted by these types of sensors. It has been especially evident in the field of volatile organic compound (VOC) detection, where traditional techniques, such as chromatography-mass spectrometry [18–21], generally lack ease of use, need specialized personnel and elaborate protocols, and are characterized by high cost of the devices and insufficient flexibility. When early warning and quality monitoring applications are requested, resistive sensors are valuable alternatives. Indeed, in some specific applications, VOCs, such as diacetyl, need to be continuously monitored [22]. During the alcoholic fermentation processes of wine, beer and distilled beverages, diacetyl is naturally produced in small quantities, and it confers a characteristic butter-like aroma [23–27]. However, high diacetyl concentration (>5 mg/L) imparts an undesirable flavor indicating an alteration in the production or storage process. The time-monitoring of diacetyl concentrations can contribute to identifying and characterizing the quality of products [22]. Moreover, during this process, the so-called alcoholic fermentation, carbon dioxide and ethanol are produced too, and the influence of the atmosphere composition was reported to be fundamental in diacetyl detection and monitoring [28]. Till now, few works reported the detection of gaseous diacetyl mainly by means of complex array sensors [27,29,30]. Itoh et al. used variously-doped SnO_2 and CeO_2 to detect diacetyl among other volatile compounds [27], whereas PPy–V_2O_5 and PPy–ZnO nanocomposite fibers were successfully used by Pirsa et al. to discriminate the presence of diacetyl in fermented products [29]. Bailey et al. used an array of conducting polymers mounted on an electronic chip to discriminate between beers and to recognize the presence of volatile compounds, such as diacetyl [30].

In this work, tin oxide (SnO_2) is used to develop a high-performance resistive MOS sensor for the detection of VOCs [17,31], diacetyl in particular, thanks to the peculiar electrical properties of its microsphere structure [13]. The detection of diacetyl is performed by sampling the head space above the liquid solutions at different diacetyl concentrations (0.2–3.2 mg/L). The sensor has been tested using both aqueous and alcoholic (5% ethanol) diacetyl solutions in different carrier/regeneration gases, able to reproduce the typical alcoholic fermentation environment, rich in carbon dioxide (CO_2) and poor of oxygen. The revealed performance shows the promising ability of the of SnO_2 sensor to discriminate between different diacetyl concentrations in anaerobic atmospheres and in the contemporary presence of ethanol, generally characterized by a strong signal, identified as the major constituent of the headspace of alcoholic beverages [32]. Response and recovery time are also considered for a better investigation.

2. Materials and Methods

2.1. Tin Oxide Preparation and Charactetrization

A facile hydrothermal procedure was used to prepare SnO_2 powder, as reported elsewhere [13]. In detail, $SnCl_2$(II), used as the tin oxide precursor, was solubilized in ethanol (7.57 g/L). In order to fully dissolve $SnCl_2$, the mixture was firstly sonicated for 20 min and then it was transferred into a Teflon-lined stainless-steel autoclave at 200 °C for 6 h and 40 min. The solution was finally cooled down to room temperature. The obtained yellow solid was collected by centrifugation and washed in ethanol until the total removal of chloride ions. After ethanol washing the sample was dried at 80 °C overnight. The sample preparation procedure is schematically depicted in Figure 1.

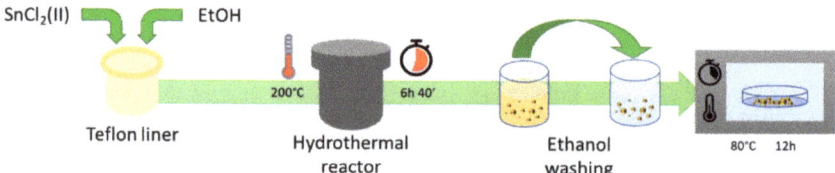

Figure 1. Synthesis procedure of SnO_2 metal oxide.

The sample was investigated using X-ray diffraction (XRD) analysis in the 2θ range from 10° to 65° (Cu Kα λ = 1.54056 Å) using steps of 0.02° and a count time of 5 s per step, with a temperature effect evaluation within 25–500 °C by using a PANalytical Empyrean diffractometer equipped with an Anton Paar heat chamber. Peaks were attributed in accordance with the Crystallography Open Database (COD). The microstructure's morphology was studied by scanning electron microscopy, SEM, (Phenom Pro X).

2.2. Sensor Preparation and Testing

The sensing film was realized using a tin oxide/ethanol mixture sonicated for 15 min, in order to obtain a homogeneous paste, and deposited on an alumina planar substrate (6 mm × 3 mm) supplied with interdigitated Pt electrodes on the front side and a heating element on the back side [33]. The paste was dried at room temperature and finally annealed at 80 °C for 1 h. Before measurements, in order to improve the deposited film stability, the sensor was conditioned in air at 400 °C for 2 h. Tests were performed by positioning the sensor in a testing cell and flowing a total gas stream of 100 sccm. To assess the different working conditions' effects on the performance of the SnO_2 sensor, different pure gas atmospheres, namely Air, N_2 and CO_2 were evaluated. Diacetyl was chosen as the target analyte, both in aqueous and in alcoholic (5% ethanol) solutions. Analyte vapor was obtained by bubbling carrier gases in the solution that was maintained at 20 °C during the tests. All gas fluxes were measured by computer-controlled mass flow meters, and humidity was monitored and controlled to oscillate between 5% and 10%. The sensors' resistance data were collected in four-point mode by an Agilent 34970A (Santa Clara, CA, USA) multimeter, while sensor temperature was controlled by using a dual-channel power supplier instrument (Agilent E3632A, Santa Clara, CA, USA).

The sensor working temperature was fixed at 200 °C. Sensor response was defined as R/R_0 when the sensor showed an n-type response and the sensor resistance R measured in the presence of the reducing analyte gas decreased in respect to the sensor resistance R_0 of the sensor exposed to the carrier flow. Differently, when the sensor showed a p-type response and the sensor resistance R in the presence of the same analyte gas increased, the response was defined as R_0/R [34]. The response and recovery time of the sensor were defined as the time needed for the sensor to reach 90% of its saturation limit after the exposure to the analyte gas and as the time needed for the sensor to reach the 10% of its original resistance value once the target gas was switched off and the sensor was exposed to the carrier gas only, respectively [14].

3. Results and Discussion

3.1. SnO_2 Powder Characterization

The crystalline microstructure of the prepared SnO_2 was studied by XRD analysis, as shown in Figure 2a. The main characteristic peaks are centered at 26.7°, 34.2°, 37.8°, and 51.9° and correspond to the (110), (101), (200), and (211) SnO_2 crystal planes, respectively. The peaks appeared broad and with weak intensity, thus resulting with small average crystallite size and low crystallinity. This latter one was ascribed to the presence of defects on the material surface, which greatly affect the reactive sensing sites and electronic structure of gas-sensing material, influencing the gas sensing properties [35]. XRD spectra acquired at increasing temperatures, from 25 °C up to 500 °C, proved the high stability of the sample, being all the registered patterns superimposable.

The morphology of the powder was studied by SEM analysis. Figure 2b shows that sample is composed of homogeneous small particles of spherical form with size distribution ranging from 0.8 to 2.1 µm. A compact oxide thin film was used as the sensing layer.

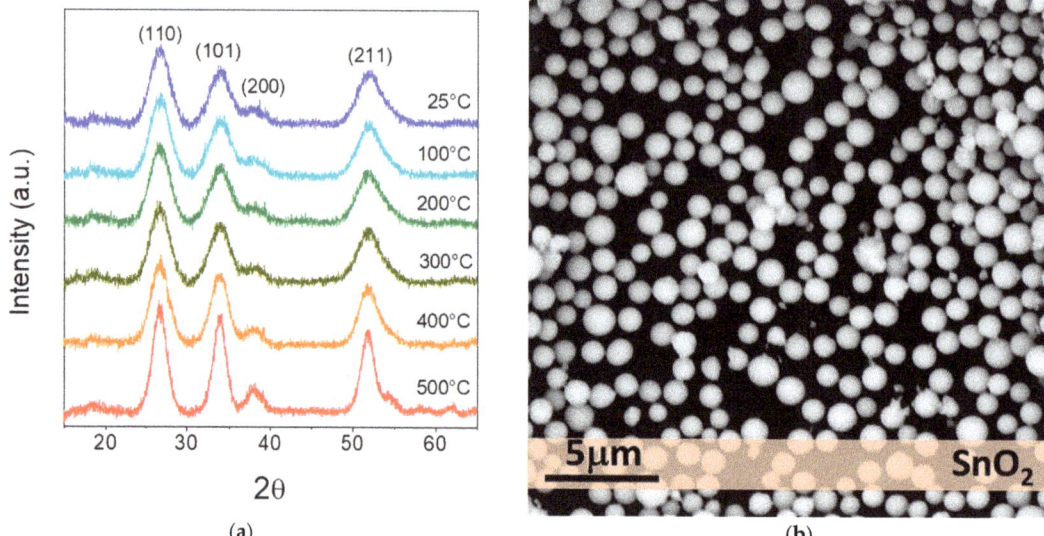

Figure 2. (a) XRD diffractograms versus temperatures and (b) SEM image of SnO_2 synthesized sample.

3.2. Gas Sensor Measurements

First, to study the behavior of the tin oxide sensor in different working atmospheres, the sensing properties toward vapors of diacetyl in aqueous solution (0.4 mg/L) were tested in oxidant, inert and reducing atmospheres, using as carrier and regeneration flow air N_2 and CO_2.

Tin oxide is well known as an MOS whose behavior is n-type. Indeed, the electrical resistance of the layer decrease when the sensor is exposed to reducing gas such as NH_3 [36], CO [37], CH_4 [38], H_2S [39] or SO_2 [40] due to the oxidation of the target gas on the SnO_2 surface. Measurements carried out under air and N_2 atmospheres (Figure 3a), confirmed the n-type behavior of the SnO_2 sensor with $R < R_0$, whereas the measurement carried out in CO_2 ambient showed an inverse behavior, with $R > R_0$, proving indeed that in CO_2, tin oxide worked as a p-type sensor.

3.2.1. Sensor Response in Air Atmosphere

SnO_2 response in air was widely studied and reported in literature [14]. Generally, in air the surface of SnO_2 is covered with negatively charged oxygen adsorbates (O_2^-, O^-, and O^{2-}). The formation of such oxygen adsorbates extracts electrons from the conduction band of SnO_2 bulk forming an electron depletion layer on the SnO_2 grains surface (space-charge region) and a potential barrier at the grain boundaries (Figure 3b) [41]; the sensor has a high resistance value (~30 MΩ). As soon as the sensor interacts with the analyte vapor, the diacetyl molecules are oxidized by oxygen species on the surface and the depleted layer electrons are fed back into SnO_2 bulk, thus a narrowed depletion layer and a reduction of the space-charge region is detected. Therefore, the sensor resistance decrease. In this case, exposing the sensor to 0.4 mg/L diacetyl solution vapors, the sensor response is equal to 1.43 and it results faster in the first seconds after the exposure, with a total response time of 3.0 min. Despite a very fast response being first observed, a delay in the sensor signal rise is detected; it demands a longer time to reach the final value. Indeed, the recording of the dynamic behavior of a sensor can be confusing, because the measured change in signal depends both on the intrinsic reaction of the gas sensing mechanism and on a delayed change in test gas concentration [42]. Moreover, it should be highlighted that response/recovery times may be dependent on the measurement procedure.

On the contrary the recovery time is constant and slow (10.2 min).

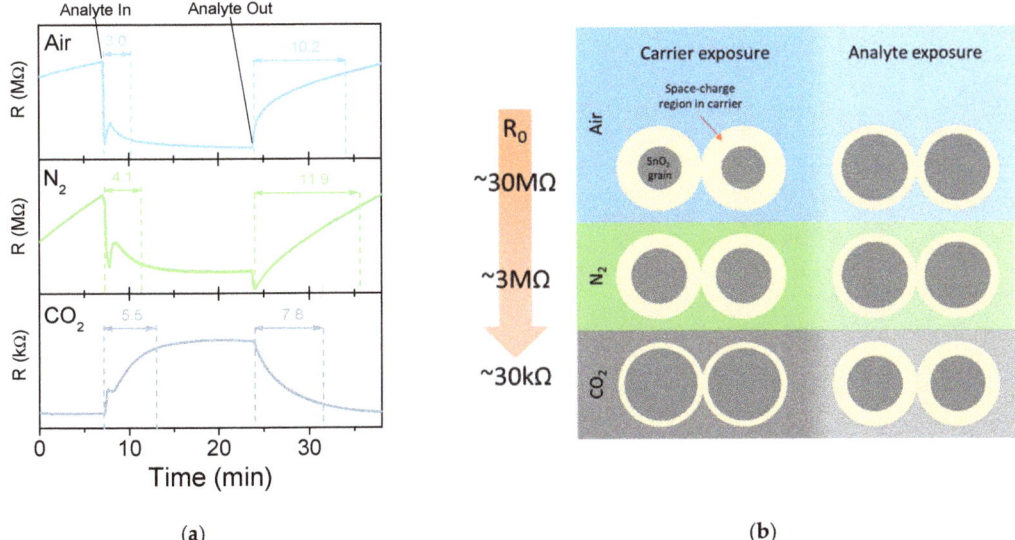

Figure 3. Sensor measurements acquired in different atmospheres: (**a**) transient sensor response of 0.4 mg/L diacetyl in aqueous solution in air, N_2 and CO_2 environments; (**b**) schematic illustration of the sensing mechanism of SnO_2 powder to diacetyl in carrier and analyte exposure.

3.2.2. Sensor Response in N_2 Atmosphere

Under N_2 atmosphere the sensor showed a lower baseline resistance than in air (~3 MΩ). This result is compatible with the reduction of the density of oxygen species bound to the surface and then the increment of free carrier electrons into the conduction band that reduce the space-charge region at SnO_2 grains boundaries. The response to 0.4 mg/L diacetyl exposure confirms a typical n-type sensing, with a resistance decrement due to the interaction of diacetyl molecules with the surface. In addition, sensors tested under N_2 conditions shows a response comparable in terms of shape and magnitude with that observed under air, even if the sensor in N_2 atmosphere do not show a complete recovery. A similar trend was already observed with a SnO_2 sensor in argon atmosphere [43]. Measurements performed in air and N_2 appeared with the same shape. Previews works indicated that despite a traditional n-type response was observed, in absence of oxygen the sensing mechanism is related to the direct adsorption of VOC on the MOS surface [43].

The sensor response in N_2 is 1.60, but both response and recovery time result longer than in air measurements cause the poor interaction between N_2 carrier and the sensor surface.

3.2.3. Sensor Response in CO_2 Atmosphere

In a pure-CO_2 atmosphere the resistance R is greater than R_0, revealing a p-type behavior. In this condition the sensor is able to detect diacetyl and shows a total recovery in short time.

In a CO_2 atmosphere R resulted smaller than both in air and N_2 (~30 kΩ). Indeed, according to Wang et al., when SnO_2 works at 240 °C in high CO_2 concentrations and less than 14% relative humidity conditions, CO_2 acts as an electron donor, as a weak reducing agent [14]. Such a response-type transition is also observed for other materials, and it is due to the inversion of the conduction type of major carriers, which limits the dynamic range of the sensor at high concentration [44]. In the pure-CO_2 atmosphere the sensor response to 0.4 mg/L diacetyl exposure is 1.13, with response and recovery times of 5.5 min and 7.8 min, respectively. Recovery time results are shorter than in air and in N_2 atmosphere cases.

3.2.4. Sensor Response towards Diacetyl in Aqueous and Alcholic Solutions

Once the effect of the different carrier/regeneration atmospheres on the performance of SnO_2 towards diacetyl was evaluated, just oxygen-deficient atmospheres were considered. Indeed, alcoholic fermentation, producing ethanol and carbon dioxide, usually occurs in anaerobic environments, since it does not require oxygen. Hence, tests were further performed in oxygen-deficient atmospheres, N_2 and CO_2, in: (i) aqueous diacetyl solution (0.4 mg/L); (ii) alcoholic (5% ethanol) diacetyl solution (0.4 mg/L), and (iii) alcoholic (5% ethanol) solution, for comparison purposes. The response values along with the response and recovery time are summarized in Figure 4.

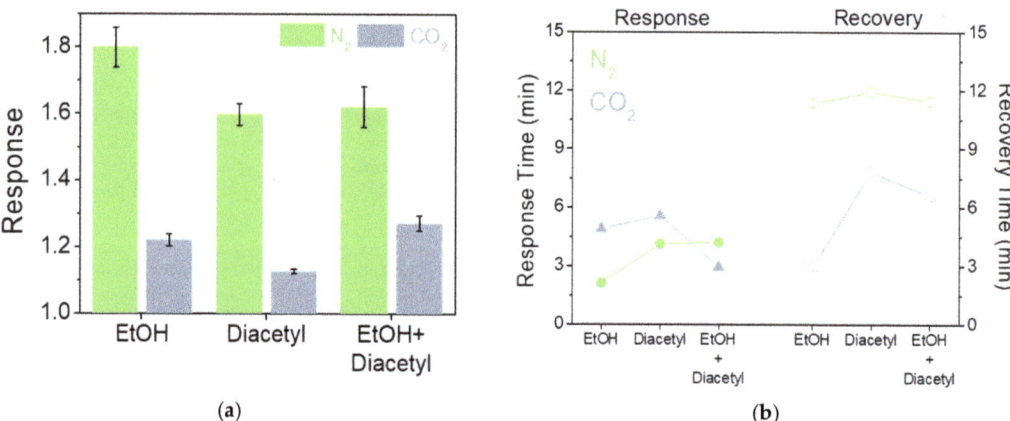

Figure 4. (a) Sensor response in N_2 and CO_2 to alcoholic (5% ethanol) solution, aqueous diacetyl solution (0.4 mg/L) and alcoholic (5% ethanol) diacetyl solution (0.4 mg/L). Error bars are calculated on three runs; (b) response and recovery times of sensor exposed at different concentration of diacetyl in aqueous and alcoholic solutions in CO_2 and N_2 ambient. Error bars are calculated on three runs.

The sensor in both N_2 and CO_2 environments shows a higher response to ethanol than diacetyl, reaching 1.80 and 1.22, respectively, with shorter response and recovery times. In alcoholic diacetyl solution, the response in N_2 ambient registers a value of 1.61, similar to the value found with the only-diacetyl solution, and response and recovery time results are similar, showing that under N_2 atmosphere it is very difficult to distinguish diacetyl effects in aqueous or alcoholic solutions. Instead, in a CO_2 atmosphere an increment in the response towards diacetyl alcoholic solution, with a final value of 1.27, is detected. In turns, the sensor, despite the lower response than in CO_2, succeeds in distinguishing diacetyl under fermentation conditions. Moreover, response time in CO_2 shows a marked decrement (46%) compared to only diacetyl. Similarly, recovery time is reduced too.

3.2.5. Effect of Diacetyl Concentrations in Alcholic and Aqueous Solutions

The fabricated SnO_2 sensor was investigated by varying the diacetyl concentration both in aqueous and alcoholic solutions in a CO_2 environment (Figure 5). The response curves for different diacetyl concentrations (0.05–3.2 mg/L) are reported in Figure 5a,b. It can be seen that the response of the SnO_2 sensor increases by increasing diacetyl concentration with an exponential trend. The response of the sensor (Y) can be fitted as a logarithmic function of Y = 1.14 + 0.07logX with X representing the diacetyl concentration in mg/L and with a regression coefficient R^2 of 0.97, as shown in the inset of Figure 5a. An exponential trend of the response depending on the diacetyl concentration in alcoholic solution is instead reported in Figure 5b. The sensor response in the alcoholic solution results higher than diacetyl in a water solution, following a logarithm profile. This is due to the high sensitivity of SnO_2 to ethanol as reported in previous works [37,38]. Moreover, the presence of ethanol markedly enhanced the gas sensing performance toward diacetyl in the alcoholic solution.

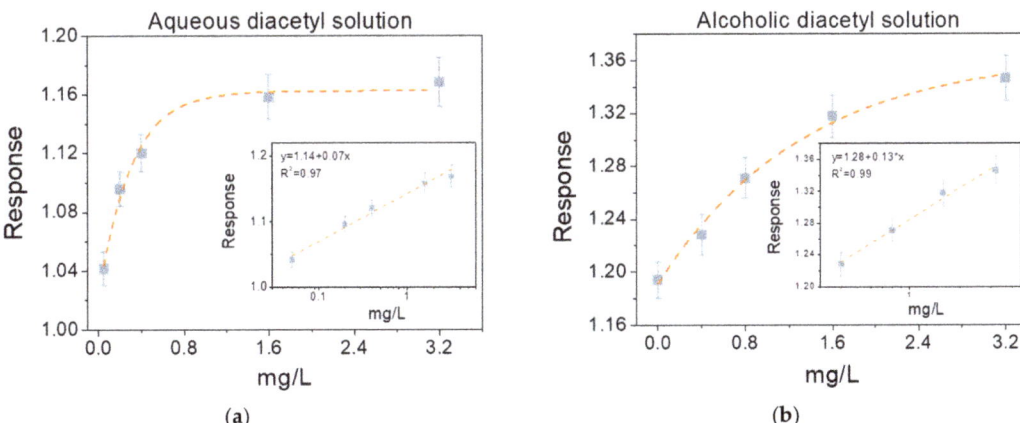

Figure 5. Sensor response in CO_2 atmospheres at different diacetyl concentrations in (**a**) aqueous solution and (**b**) alcoholic solution. Error bars are calculated on three runs.

By fitting the sensor response with a logarithmic function Y = 1.28 + 0.13logX (R^2 = 0.99) it is evident how the slope of the curve is greater than the response of diacetyl in water solution. From the fit it is possible to extract the lower detection limit (LDL), the minimum concentration of gas detectable by gas sensor [45,46].

For SnO_2 sensor LDL can be estimated by the extrapolation of the linear fit of the response curve down to the minimum response R_{min}/R_0 where $R_{min} = R_0 - 3\sigma$, and σ is the standard deviation of the baseline resistance before analyte exposure. For diacetyl in an aqueous solution LDL results 0.1 mg/L, while for measurements performed on diacetyl in an alcoholic solution LDL is less than 0.01 mg/L. In turn, ethanol presence increases the sensitivity of the sensor, proving the capability of the SnO_2 sensor to detect diacetyl in alcoholic solution in a CO_2 atmosphere.

3.2.6. Response and Recovery Time in Alcholic Diacetyl Solution

Response and recovery times of diacetyl in alcoholic solution under a CO_2 atmosphere are analyzed, being important to understand the performance of the sensor.

Figure 6a shows response/recovery times of the SnO_2 sensor in different diacetyl concentrations. Their analysis shows that for all the evaluated diacetyl concentrations, recovery time is higher than response time. Response and recovery time values decrease with an exponential law upon increasing the concentration; response time is almost halved, from 3.2 min to 1.7 min with a diacetyl increment from 0.4 mg/L to 3.2 mg/L. Analogously, the recovery time decreases from 6.6 min to 2.2 min in the same working conditions. According to literature data [42], basic effects of surface covering kinetics and diffusion may be the cause of a dependence on concentration change. Moreover, since the semiconductor can be considered as an RC filter with a time constant τ, electronic effects may have influence too. Due to the increasing of diacetyl concentration, the resistance decreases and thus also the time constant.

Indeed, the sensor resistance as a function of time during the exposure to the analyte and during the recovery can be described through the following equations [42]:

$$R(t) = R_{max} - (R_{max} - R_0)\exp\left[\frac{-t}{\tau_{resp}}\right], \quad (1)$$

$$R(t) = R_0 + (R_{max} - R_0)\exp\left[\frac{-t}{\tau_{rec}}\right], \quad (2)$$

where $R(t)$ is the resistance after the analyte exposure, R_0 is a constant, R_{max} is the saturated resistance, t is the time, and τ_{resp} and τ_{rec} are the response and recovery time constants, respectively. Figure 6b shows the plot of time constants as a function of diacetyl concentration. The curves show that the time constants decreased as diacetyl concentration increased. According to response and recovery time analysis, the range of τ values recorded at low diacetyl concentration is much wider than that recorded at high diacetyl concentration, where τ is less than 50 s.

The parameter τ is therefore considered a more suitable parameter to characterize sensors response and recovery times.

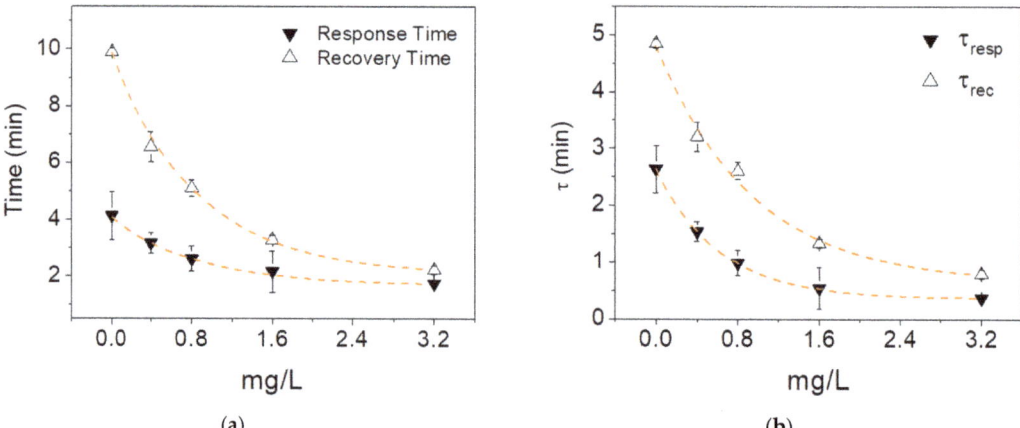

Figure 6. Time parameters for sensor exposed at different concentration of diacetyl in alcoholic solution and a CO_2 environment; (**a**) sensor response and recovery times at different diacetyl concentration; (**b**) plot of time constants as a function of diacetyl concentration. Error bars are calculated on three runs.

4. Conclusions

A SnO_2-based MOS sensor is used to investigate the detection of VOCs in fermented beverages such as beer, wine and distillates. Synthesized and characterized SnO_2 is used to discriminate the concentration of diacetyl vapors in different operating atmospheres, as to replicate a typical alcoholic fermentation scenario.

Under air, N_2 and CO_2 gases, SnO_2 shows different behavior and baseline resistances due to different interactions with its surface. In particular, measurements carried out under air and N_2 atmospheres show the n-type behavior of the SnO_2 sensor, whereas measurements carried out in CO_2 ambient prove that SnO_2 acted as a p-type sensor.

Tests performed in aqueous and alcoholic diacetyl solutions show good response in terms of diacetyl detection in the range of concentration 0.2–3.2 mg/L at 200 °C in CO_2 atmosphere, with LDL of 0.1 mg/L and 0.01 mg/L, respectively.

Response and recovery times trends reveal the diacetyl concentration dependance, with response and recovery times constant less than 50 s.

Author Contributions: Conceptualization, A.G., A.F., L.B. and A.D.; methodology, A.G., A.F. and L.B.; formal analysis, A.G. and A.F.; investigation, A.G. and A.F.; data curation, A.G.; writing—original draft preparation, A.G. and A.F.; writing—review and editing, A.G. and L.B.; supervision L.B. and A.D. All authors have read and agreed to the published version of the manuscript.

Funding: This research was co-financed with the support of the European Commission, the European Social Fund and the Calabria Region. The author is solely responsible for this research and the European Commission and the Calabria Region decline any responsibility for the use that may be made of the information contained therein.

Institutional Review Board Statement: Not applicable.

Conflicts of Interest: The authors declare no conflict of interest.

References

1. Wu, H.; Shi, L. Real-time anomaly detection in gas sensor streaming data. *Int. J. Embed. Syst.* **2021**, *14*, 81–88. [CrossRef]
2. Nazemi, H.; Joseph, A.; Park, J.; Emadi, A. Advanced Micro- and Nano-Gas Sensor Technology: A Review. *Sensors* **2019**, *19*, 1285. [CrossRef]
3. Malara, A.; Bonaccorsi, L.; Donato, A.; Frontera, P.; Neri, G. Doped Zinc Oxide Sensors for Hexanal Detection. In *Lecture Notes in Electrical Engineering*; Springer: Singapore, 2020; pp. 279–285.
4. Wang, Z.; Zhu, L.; Sun, S.; Wang, J.; Yan, W. One-Dimensional Nanomaterials in Resistive Gas Sensor: From Material Design to Application. *Chemosensors* **2021**, *9*, 198. [CrossRef]
5. Zhou, X.; Lee, S.; Xu, Z.; Yoon, J. Recent Progress on the Development of Chemosensors for Gases. *Chem. Rev.* **2015**, *115*, 7944–8000. [CrossRef] [PubMed]
6. Faggio, G.; Gnisci, A.; Messina, G.D.S.; Lisi, N.; Capasso, A.; Lee, G.-H.; Armano, A.; Sciortino, A.; Messina, F.; Cannas, M.; et al. Carbon Dots Dispersed on Graphene/SiO2/Si: A Morphological Study. *Phys. Status Solidi* **2019**, *216*, 1800559. [CrossRef]
7. Sawalha, S.; Moulaee, K.; Nocito, G.; Silvestri, A.; Petralia, S.; Prato, M.; Bettini, S.; Valli, L.; Conoci, S.; Neri, G. Carbon-dots conductometric sensor for high performance gas sensing. *Carbon Trends* **2021**, *5*, 100105. [CrossRef]
8. Malara, A.; Leonardi, S.G.; Donavita, A.; Fazio, E.; Stelitano, S.; Neri, G.; Neri, F.; Santangelo, S. Origin of the different behavior of some platinum decorated nanocarbons towards the electrochemical oxidation of hydrogen peroxide. *Mater. Chem. Phys.* **2016**, *184*, 269–278. [CrossRef]
9. Nag, A.; Mitra, A.; Mukhopadhyay, S. Graphene and its sensor-based applications: A review. *Sens. Actuators A Phys.* **2018**, *270*, 177–194. [CrossRef]
10. Busacca, C.; Donato, A.; Faro, M.L.; Malara, A.; Neri, G.; Trocino, S. CO gas sensing performance of electrospun Co_3O_4 nanostructures at low operating temperature. *Sens. Actuators B Chem.* **2020**, *303*, 127193. [CrossRef]
11. Kailasa, S.; Reddy, M.S.B.; Maurya, M.R.; Rani, B.G.; Rao, K.V.; Sadasivuni, K.K. Electrospun Nanofibers: Materials, Synthesis Parameters, and Their Role in Sensing Applications. *Macromol. Mater. Eng.* **2021**, *306*, 2100410. [CrossRef]
12. Radhakrishnan, J.; Kumara, M. Geetika Effect of temperature modulation, on the gas sensing characteristics of ZnO nanostructures, for gases O_2, CO and CO_2. *Sens. Int.* **2021**, *2*, 100059. [CrossRef]
13. Zhang, L.; Tong, R.; Ge, W.; Guo, R.; Shirsath, S.E.; Zhu, J. Facile one-step hydrothermal synthesis of SnO_2 microspheres with oxygen vacancies for superior ethanol sensor. *J. Alloys Compd.* **2020**, *814*, 152266. [CrossRef]
14. Wang, D.; Chen, Y.; Liu, Z.; Li, L.; Shi, C.; Qin, H.; Hu, J. CO2-sensing properties and mechanism of nano-SnO_2 thick-film sensor. *Sens. Actuators B Chem.* **2016**, *227*, 73–84. [CrossRef]
15. Hoa, T.T.N.; Van Duy, N.; Hung, C.M.; Van Hieu, N.; Hau, H.H.; Hoa, N.D. Dip-coating decoration of Ag_2O nanoparticles on SnO_2 nanowires for high-performance H2S gas sensors. *RSC Adv.* **2020**, *10*, 17713–17723. [CrossRef]
16. Sun, P.; Mei, X.; Cai, Y.; Ma, J.; Sun, Y.; Liang, X.; Liu, F.; Lu, G. Synthesis and gas sensing properties of hierarchical SnO_2 nanostructures. *Sens. Actuators B Chem.* **2013**, *187*, 301–307. [CrossRef]
17. Gnisci, A.; Fotia, A.; Malara, A.; Bonaccorsi, L.; Frontera, P.; Donato, A. SnO_2 sensing performance toward volatile organic compounds. In Proceedings of the CSAC2021: 1st International Electronic Conference on Chemical Sensors and Analytical Chemistry, Online, 1–15 July 2021; Malara, A., Ed.; MDPI: Basel, Switzerland, 2021.
18. Parish, M.E.; Braddock, R.J.; Wicker, L. Gas Chromatographic Detection of Diacetyl in Orange Juice. *J. Food Qual.* **1990**, *13*, 249–258. [CrossRef]
19. Macciola, V.; Candela, G.; De Leonardis, A. Rapid gas-chromatographic method for the determination of diacetyl in milk, fermented milk and butter. *Food Control.* **2008**, *19*, 873–878. [CrossRef]
20. Otsuka, M.; Ohmori, S. Simple and sensitive determination of diacetyl and acetoin in biological samples and alcoholic drinks by gas chromatography with electron-capture detection. *J. Chromatogr. B Biomed. Sci. Appl.* **1992**, *577*, 215–220. [CrossRef]
21. Tian, J. Determination of several flavours in beer with headspace sampling–gas chromatography. *Food Chem.* **2010**, *123*, 1318–1321. [CrossRef]
22. Clark, S.; Winter, C.K. Diacetyl in Foods: A Review of Safety and Sensory Characteristics. *Compr. Rev. Food Sci. Food Saf.* **2015**, *14*, 634–643. [CrossRef]

23. Li, P.; Guo, X.; Shi, T.; Hu, Z.; Chen, Y.; Du, L.; Xiao, D. Reducing diacetyl production of wine by overexpressing BDH1 and BDH2 in Saccharomyces uvarum. *J. Ind. Microbiol. Biotechnol.* **2017**, *44*, 1541–1550. [CrossRef]
24. Martineau, B.; Acree, T.E.; Henick-Kling, T. Effect of wine type on the detection threshold for diacetyl. *Food Res. Int.* **1995**, *28*, 139–143. [CrossRef]
25. Krogerus, K.; Gibson, B.R. 125thAnniversary Review: Diacetyl and its control during brewery fermentation. *J. Inst. Brew.* **2013**, *119*, 86–97. [CrossRef]
26. Lee, H.-H.; Lee, K.-T.; Shin, J.-A. Analytical method validation and monitoring of diacetyl in liquors from Korean market. *Food Sci. Biotechnol.* **2017**, *26*, 893–899. [CrossRef] [PubMed]
27. Itoh, T.; Koyama, Y.; Shin, W.; Akamatsu, T.; Tsuruta, A.; Masuda, Y.; Uchiyama, K. Selective Detection of Target Volatile Organic Compounds in Contaminated Air Using Sensor Array with Machine Learning: Aging Notes and Mold Smells in Simulated Automobile Interior Contaminant Gases. *Sensors* **2020**, *20*, 2687. [CrossRef] [PubMed]
28. Portno, A.D. The influence of oxygen on the production of diacetyl during fermentation and conditioning. *J. Inst. Brew.* **1966**, *72*, 458–461. [CrossRef]
29. Pirsa, S.; Nejad, F.M. Simultaneous analysis of some volatile compounds in food samples by array gas sensors based on polypyrrole nano-composites. *Sens. Rev.* **2017**, *37*, 155–164. [CrossRef]
30. Bailey, T.P.; Hammond, R.V.; Persaud, K.C. Applications for an Electronic Aroma Detector in the Analysis of Beer and Raw Materials. *J. Am. Soc. Brew. Chem.* **1995**, *53*, 39–42. [CrossRef]
31. Malara, A.; Bonaccorsi, L.; Donato, A.; Frontera, P.; Piscopo, A.; Poiana, M.; Leonardi, S.G.; Neri, G. Sensing Properties of Indium, Tin and Zinc Oxides for Hexanal Detection. In *Lecture Notes in Electrical Engineering*; Springer: Singapore, 2019; pp. 39–44.
32. Ragazzo-Sanchez, J.A.; Chalier, P.; Chevalier, D.; Ghommidh, C. Electronic nose discrimination of aroma compounds in alcoholised solutions. *Sens. Actuators B Chem.* **2006**, *114*, 665–673. [CrossRef]
33. Bonaccorsi, L.; Malara, A.; Donato, A.; Donato, N.; Leonardi, S.G.; Neri, G. Effects of UV Irradiation on the Sensing Properties of In_2O_3 for CO Detection at Low Temperature. *Micromachines* **2019**, *10*, 338. [CrossRef]
34. Williams, D.E. Semiconducting oxides as gas-sensitive resistors. *Sens. Actuators B Chem.* **1999**, *57*, 1–16. [CrossRef]
35. Liu, L.; Shu, S.; Zhang, G.; Liu, S. Highly Selective Sensing of C2H6O, HCHO, and C3H6O Gases by Controlling SnO_2 Nanoparticle Vacancies. *ACS Appl. Nano Mater.* **2018**, *1*, 31–37. [CrossRef]
36. Shahabuddin, M.; Sharma, A.; Kumar, J.; Tomar, M.; Umar, A.; Gupta, V. Metal clusters activated SnO_2 thin film for low level detection of NH3 gas. *Sens. Actuators B Chem.* **2014**, *194*, 410–418. [CrossRef]
37. Hermida, I.D.P.; Wiranto, G.; Hiskia; Nopriyanti, R. Fabrication of SnO_2 based CO gas sensor device using thick film technology. *J. Phys. Conf. Ser.* **2016**, *776*, 012061. [CrossRef]
38. Sedghi, S.M.; Mortazavi, Y.; Khodadadi, A.A. Low temperature CO and CH4 dual selective gas sensor using SnO_2 quantum dots prepared by sonochemical method. *Sens. Actuators B Chem.* **2010**, *145*, 7–12. [CrossRef]
39. Devi, G.S.; Manorama, S.; Rao, V. High sensitivity and selectivity of an SnO_2 sensor to H2S at around 100 °C. *Sens. Actuators B Chem.* **1995**, *28*, 31–37. [CrossRef]
40. Das, S.; Girija, K.; Debnath, A.; Vatsa, R. Enhanced NO_2 and SO_2 sensor response under ambient conditions by polyol synthesized Ni doped SnO_2 nanoparticles. *J. Alloys Compd.* **2021**, *854*, 157276. [CrossRef]
41. Shimizu, Y. *SnO2 Gas Sensor BT—Encyclopedia of Applied Electrochemistry*; Kreysa, G., Ota, K., Savinell, R.F., Eds.; Springer: New York, NY, USA, 2014; pp. 1974–1982. ISBN 978-1-4419-6996-5.
42. Hübert, T.; Majewski, J.; Banach, U.; Detjens, M.; Tiebe, C. Response Time Measurement of Hydrogen Sensors. 2017. Available online: https://hysafe.info/uploads/2017_papers/211.pdf (accessed on 28 December 2021).
43. Abokifa, A.A.; Haddad, K.; Fortner, J.; Lo, C.S.; Biswas, P. Sensing mechanism of ethanol and acetone at room temperature by SnO_2 nano-columns synthesized by aerosol routes: Theoretical calculations compared to experimental results. *J. Mater. Chem. A* **2018**, *6*, 2053–2066. [CrossRef]
44. Liu, X.-L.; Zhao, Y.; Ma, S.-X.; Zhu, S.-W.; Ning, X.-J.; Zhao, L.; Zhuang, J. Rapid and Wide-Range Detection of NOx Gas by N-Hyperdoped Silicon with the Assistance of a Photovoltaic Self-Powered Sensing Mode. *ACS Sens.* **2019**, *4*, 3056–3065. [CrossRef]
45. Barsan, N.; Weimar, U. Conduction Model of Metal Oxide Gas Sensors. *J. Electroceram.* **2001**, *7*, 143–167. [CrossRef]
46. Santos, G.T.; Felix, A.A.; Orlandi, M.O. Ultrafast Growth of h-MoO_3 Microrods and Its Acetone Sensing Performance. *Surfaces* **2021**, *4*, 9–16. [CrossRef]

Article

Morphological Observation of LiCl Deliquescence in PDMS-Based Composite Foams

Emanuela Mastronardo [1,*], Elpida Piperopoulos [1,2], Davide Palamara [1], Andrea Frazzica [2] and Luigi Calabrese [1,2,*]

1 Engineering Department, University of Messina, 98166 Messina, Italy; epiperopoulos@unime.it (E.P.); davide.palamara@unime.it (D.P.)
2 CNR ITAE, 98126 Messina, Italy; andrea.frazzica@itae.cnr.it
* Correspondence: emastronardo@unime.it (E.M.); luigi.calabrese@unime.it (L.C.)

Abstract: The LiCl-based heat storage system exhibits a high-energy density, making it an attractive and one of the most investigated candidates for low-temperature heat storage applications. Nevertheless, lithium chloride, due to its hygroscopic nature, incurs the phenomenon of deliquescence, which causes some operational challenges, such as agglomeration, corrosion, and swelling problems during hydration/dehydration cycles. Here, we propose a composite material based on silicone vapor-permeable foam filled with the salt hydrate, hereafter named LiCl-PDMS, aiming at confining the salt in a matrix to prevent deliquescence-related issues but without inhibiting the vapour flow. In particular, the structural and morphological modification during hydration/dehydration cycles is investigated on the composite foam, which is prepared with a salt content of 40 wt.%. A characterization protocol coupling temperature scanned X-ray diffraction (XRD) and environmental scanning electron microscopy (ESEM) analysis is established. The operando conditions of the dehydration/hydration cycle were reproduced while structural and morphological characterizations were performed, allowing for the evaluation of the interaction between the salt and the water vapor environment in the confined silicon matrix. The material energy density was also measured with a customized coupled thermogravimetric/differential scanning calorimetric analysis (TG/DSC). The results show an effective embedding of the material, which limits the salt solution release when overhydrated. Additionally, the flexibility of the matrix allows for the volume shrinkage/expansion of the salt caused by the cyclic dehydration/hydration reactions without any damages to the foam structure. The LiCl-PDMS foam has an energy density of 1854 kJ/kg or 323 kWh/m^3, thus making it a competitive candidate among other LiCl salt hydrate composites.

Keywords: lithium chloride hydrate; composite foam; deliquescence; thermochemical storage; in situ characterization

1. Introduction

The energy transition is a necessary path toward the transformation of the worldwide energy sector from fossil-based to zero carbon by 2050, which is fundamental to reduce energy-related CO$_2$ emissions to limit climate change [1]. One of the main pillars for the energy transition is the global spread of the use of solar energy, being basically endless and free [2]. However, its diurnal nature, weather conditions, and seasonality might make solar energy unavailable when required. This time lag between energy supply and demand is a critical point to be addressed for renewable solar energy development and replacement of fossil-based energy. For this aim, thermal energy storage (TES) is an efficient and effective means to store solar energy when in excess, thus shifting the peak load demand into off-peak hours.

Among TES technologies, the thermochemical one exploits materials with a large reaction enthalpy and reversibility. Inorganic salt hydrate (M$_n$A$_m$·XH$_2$O) is a widely used class and one of the most promising investigated classes of materials for thermochemical heat storage [3,4], which is based on the following storage reaction:

$$M_nA_m \cdot nH_2O + heat \leftrightarrow M_nA_m \cdot (n-x)H_2O + xH_2O$$

Heat is transferred from a selected source to the material and the dehydration reaction takes place (storage step). The heat is stored for as long as the salt is in the dehydrated form. When heat is required, water is made accessible to the salt, the reversible hydration takes place, and heat is released (release step). One of the main advantages of this technology is, indeed, the possibility to control the heat release by controlling the water vapor accessibility to the dehydrated salt, thus making the heat discharge available and controllable on-demand [5].

Due to this great potential, several pure salt hydrates and composite systems with different salts were investigated: e.g., LiBr [6], $MgSO_4$ [7], $CaCl_2$ [8], $SrCl_2$ [9], and LiCl [10]. Among them, LiCl is one of the salts that raised more interest for low/medium temperature (i.e., below 120 °C) TES application [11]. Indeed, the high-energy density of the LiCl-based system makes it an attractive candidate from the heat storage capacity cost perspective, thus resulting in one of the materials with a lower cost per kWh stored (~6 €/kWh) [12]. Nevertheless, lithium chloride is one of the most hygroscopic salts known and it incurs the phenomenon of deliquescence.

Deliquescence is a first-order phase transformation of the solid to a saturated solution, which occurs at a specific relative humidity (RH) inherent to the properties of the solid and the temperature [13]. When this RH is reached, the aqueous solution is the thermodynamically favored phase and dissolution begins, whereby liquid patterns or patches of solution form on the surface. As the humidity further increases, these patches merge and form a thin liquid film, which gradually thickens; the entire solid particle dissolves and transforms into a solution drop; and its radius abruptly grows. In a closed environment, a saturated solution of lithium chloride will form an equilibrium at a relative humidity of about 12.4% (20 °C). The reverse process to deliquescence is called efflorescence. Efflorescence is the process of the crystallization and expulsion of water from the crystallized material when the decreasing humidity reaches another threshold value, called efflorescence relative humidity (ERH).

Several issues arise due to the deliquescence phenomenon in TES technologies. The liquid film that forms on the surface of the salt crystal inhibits the rehydration reaction (e.g., in the case of the LiCl, LiBr, and $CaCl_2$) [14,15]. In addition, the sorbate mass transfer into the system will be hindered, causing issues such as high-pressure drops and ultimately system failure, but also corrosion issues arise due to the dripping of the salt solution to other metal components of the systems.

Accordingly, several efforts were carried out to implement LiCl-based TES technologies and, more specifically, from the material perspective. The general idea is to embed the salt into porous matrices (such as carbon foams, expanded natural graphite, zeolite, vermiculite, silica gel, etc.) to prevent deliquescence and improve water transport into the materials [8,16,17].

Zhang et al. developed a LiCl-based composite on mesoporous alumina [18]. The material showed a TES density of up to 1040 J/g. However, due to the low pore volume of the alumina, only ~15 wt.% of the salt could be loaded in the matrix, thus limiting the possibility of further increasing TES density. Multi-walled carbon nanotubes (MWCNTs), due to their extremely large and variable pore volume, were used as a matrix to host LiCl salt [19]. This composite, with a 44 wt.% content of LiCl salt, achieved a heat storage capacity of 1.6–1.7 kJ/g. More recently, a LiCl/vermiculite, with ~59 wt.% of LiCl, was developed by Grekova and coworkers [20]. The heat storage capacity of this composite is 1.8–2.6 kJ/g (224–253 Wh/m^3). It has to be pointed out that, intrinsically, the porous structure of natural mineral vermiculite is non-reproducible and so is the composite. Additionally, vermiculite suffers from hydrothermal structural stability issues that are critical for the technology's long-term durability. Yu et al. developed silica gel-LiCl composites by varying the amount of embedded salt between 10 and 40 wt.% [21]. An optimum compromise between the heat storage capacity and material stability was identified in the composition with 30 wt.% of embedded salt. For this composition, a heat storage capacity of 480 J/g was determined. The presence of salt deposited onto the external surface of the composite has clearly a detrimental effect. This issue was addressed specifically on mesoporous silica gel-LiCl composites with a post-treatment based on an adsorption phase followed by a slow desorption phase, which allowed the salt to move from the surface into the pores, likely due to a capillary effect [22].

The investigated matrices, being rigid structures, could limit the mechanical stability of the material over cycling due to the consecutive expansion and contraction of the salt grains during the hydration and dehydration steps. In this regard, the use of flexible polymeric macro-porous foam as matrices for salts was proposed in literature for other types of salts, e.g., $MgSO_4 \cdot 7H_2O$ [23,24]. The flexibility of these foams can accommodate the expansion and shrinkage of the salt hydrate volume during the hydration and dehydration reactions, thus enhancing the mechanical stability of the composite over the cycles. In addition, the foam should embed the salt, thus retaining the deliquescence that inevitably occurs due to the operating conditions.

In situ studies on TES materials can certainly provide an in-depth understanding of the reaction mechanism and materials changes during the charging/discharging operations, thus leading to valuable information for the advancement of TES materials. Here, we investigate the structural and morphological modification during hydration/dehydration cycles on a LiCl-silicon foam (LiCl-PDMS) with a salt content of 40 wt.%. Specifically, a characterization protocol coupling temperature scanned X-ray diffraction (XRD) and environmental scanning electron microscopy (ESEM) analysis was proposed. The operando conditions of the dehydration/hydration cycle were reproduced while structural and morphological characterizations were performed, allowing for the evaluation of the interaction between the salt and the water vapor environment in the confined silicon matrix.

2. Materials and Methods

2.1. LiCl-PDMS Foam Synthesis and Characterization

For this study, a silicone foam based on a poly(methylhydrosiloxane) (PMHS) and silanol-terminated polydimethylsiloxane (PDMS) mixture, filled with 40 wt.%. of the LiCl salt final weight, was used. The macro-porous composite foam was prepared by a dehydrogenative coupling reaction between hydroxyl functional materials and hydride functional siloxanes in the presence of a metal salt catalyst (bubbling agent), inducing the hydrogen evolution according to the procedure reported by Calabrese et al. [25]. Briefly, the PDMS and PMHS monomers (supplied by Gelest Inc., Morrisville, PA, USA) used as a monomer and hardener, respectively, were mixed together (PDMS:PMHS = 2:1). Then, $LiCl \cdot H_2O$ salt (>99.9%, Sigma Aldrich, St. Louis, MO, USA) was added under vigorous mixing until a homogenously and well-dispersed slurry was obtained. Anhydrous denatured ethanol (Sigma Aldrich, St. Louis, MO, USA) was added (~8 wt.%) to better mix and homogenize the slurry. A tin salt catalyst (bis(2-ethylhexanoate)tin, Gelest Inc., Morrisville, PA, USA) was added (~8 wt.%) in order to activate the reaction. Finally, the foaming process was carried out by placing the slurry in a cylindrical mold at 60 °C for 24 h and the dehydrogenative coupling reaction between the slurry constituents took place. Specifically, the hydroxyl and hydride functional groups in PDMS and PMHS, respectively, reacted, forming a siloxane link, namely Si-O-Si, that gradually led to a tri-dimensional rubber-like silicone network. Hydrogen is also a reaction by-product that acts as a foaming agent.

The homogeneity and void distribution of the LiCl-PDMS foam were analyzed on the material cross-sectional area at 50× magnification by using a 3D optical digital microscope, specifically HK-8700 (Hirox, Tokyo, Japan).

2.2. Hydration/Dehydration Cycle through Thermogravimetric Dynamic Vapor Sorption System

A complete hydration/dehydration cycle in a controlled (temperature, RH) and measurable (mass change) environment was performed through a thermogravimetric dynamic vapor sorption system (DVS Vacuum Surface Measurement Systems). The system consists of a micro-balance (precision of ±0.1 µg) and a water vapor pressure flow controller placed in the measuring chamber. Before the test, the sample was dehydrated at 150 °C under vacuum for 2 h. The hydration/dehydration cycle was performed in isothermal mode at 30 °C and varying the RH from 0 to 90%.

2.3. In-Situ Characterization of LiCl-PDMS Foam

In situ X-ray diffraction (XRD, D8 Advance Bruker diffractometer Bragg-Brentano theta-2theta configuration, Cu Kα, 40 V, 40 mA) was carried out on the LiCl-PDMS foam while dehydrating. The diffractometer was equipped with a heating chamber (HTK 1200N, Anton Paar) that enabled the material dehydration reaction from r.T. up to 80 °C under a nitrogen flow. A heating/cooling rate of 10°C/min was used. Each scan was acquired in the 2θ range of 10–80°, with a step size of 0.010° in 0.1 s, immediately after reaching the target temperature and after 30 min of holding time to allow for material equilibration. After the heating step (dehydration), the material was cooled to r.T. under nitrogen flow and XRD patterns were collected immediately after reaching the set temperature as well as after 30 min of holding time. Additionally, after cooling, the material was left for 1 h under atmospheric conditions to rehydrate and XRD was collected. Lattice parameters and the phase fraction of hydrate and anhydrous LiCl salt were evaluated by Rietveld refinement using the GSAS-II Crystallography Data Analysis Software [26].

Morphological observation of LiCl-PDMS foam while hydrating was carried out by an environmental scanning electron microscope (ESEM, FEI Quanta 450) operating with an accelerating voltage of 8 kV. Initially, the material was dehydrated in oven at 80 °C for 12 h and then placed in the ESM chamber for acquiring the micrographs under controlled water vapor atmosphere. The relative humidity (RH) was varied between 0 and 90% by tuning the temperature and water vapor pressure in the ranges of 5–40 °C and 10–800 Pa. Specifically, the micrographs were acquired after an equilibration time of 30 min under isothermal conditions (40 °C), varying the water vapor pressure from 10 Pa to 800 Pa (0.1–10.9% of relative humidity), and then under isobaric conditions (800 Pa), varying the chamber temperature from 40 °C to 5 °C (10.9–91.3% of relative humidity). The analysis was concluded, restoring the initial conditions. For clarity, a scheme of the ESEM analysis cycle conditions is shown in Figure 1.

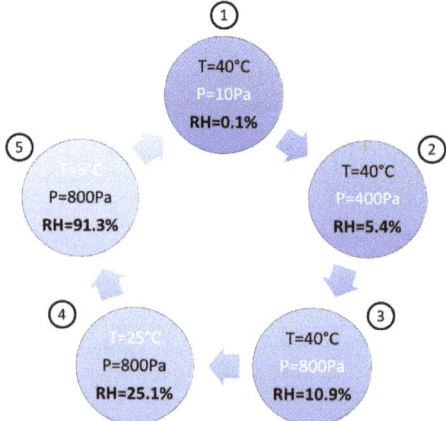

Figure 1. Schematic representation of environmental scanning electron microscopy (ESEM) analysis cycle conditions.

2.4. Hydration Heat Capacity Measurement

The hydration enthalpy, namely the heat release capacity, was evaluated through a modified coupled TG/DSC apparatus (Setaram LabsysEvo) that enables measurements under saturated vapour working conditions, as described elsewhere in literature [27,28]. Briefly, the system was equipped with a glass evaporator whose temperature was controlled by an external thermo-cryostat. The entire system was placed in a thermostatic box that allows for preventing condensation on the internal surfaces of the circuit or in the measuring chamber, and the chamber and evaporator pressures were continuously monitored. According to the testing procedure, ~16 mg of the material was placed into the measuring chamber of the TG/DSC apparatus and degassed at 150 °C for 12 h under vacuum (~10−3 mbar) to completely dehydrate the sample. Subsequently, the sample temperature was cooled down to the initial adsorption temperature (80 °C) and water vapor (vapor pressure of 12 mbar) was streamed in the measuring chamber. The sample temperature was decreased to the final discharging temperature of 35 °C, corresponding to an RH of ~22%, after which it was isothermally held for 140 min. This results in a sample mass gain due to the water uptake. Hence, the heat involved in the hydration reaction can be estimated from the integration of the DSC signal.

3. Results

3.1. LiCl-PDMS Foam Optical Analysis

The as-synthesized LiCl-PDMS foam has a measured apparent density (ρ_{foam}) and solid density (ρ_{solid}) of 0.628 and 1.094 g/cm^3, respectively, and thus a porosity of 42%, calculated as:

$$P(\%) = 1 - \frac{\rho_{foam}}{\rho_{solid}} \cdot 100 \qquad (1)$$

The apparent density was calculated as the sample weight to volume ratio, while the solid density was calculated, applying the mixture rule, by using the constituent content in the composite foam. A 3D optical image of the cross-sectional area of the LiCl-PDMS foam at 50× magnification is shown in Figure 2. The foam exhibited both open and closed porosities caused by the foaming process occurring during the synthesis. The porosities' size broadly varied from 1 to 50 mm and it can be argued that the larger porosities were likely formed due to the coalescence of several bubbles. The LiCl salt appeared to be distributed on the surface of the open porosities but was also embedded by the matrix. The optical analysis was carried out under environmental conditions and this allowed the salt to absorb the humidity present in the atmosphere and hydrate. From the inset of Figure 2, some drops are clearly visible on the surface of the foam, likely due to the deliquescence phenomenon. Indeed, the open porosities present on the surface of the foam allow for the exposure of LiCl to the humidity, thus causing the deliquescence phenomenon.

3.2. In-Situ X-ray Diffraction of LiCl-PDMS Foam during Dehydration Reaction

We report, in Figure 3, the crystal structure evolution of LiCl-PDMS foam while dehydrating through in situ XRD measurements. The XRD patterns code, the experimental conditions, the refined lattice parameters, and the volume and phase fraction are reported in Table 1.

It is well known from literature that LiCl, besides anhydrous, exists in another four solid hydrate forms with one, two, three, and five water molecules, and that these salts are extremely hygroscopic and soluble in water. As inferred from XRD analysis in Figure 2 (pattern (a)), at the initial conditions, namely r.T. and when exposed to air, the LiCl in the PDMS foam is completely amorphous (Figure 3 (pattern [a])), thus indicating that the salt is overhydrated.

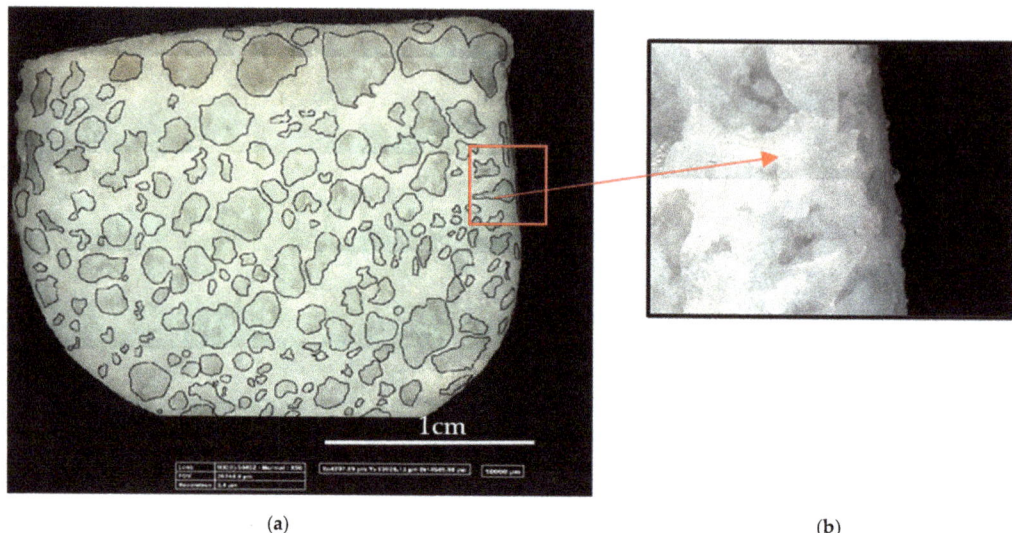

Figure 2. Three-dimensional optical image of (**a**) cross-section area of LiCl-PDMS foam at 50×
magnification. (**b**) Higher magnification of a portion of the foam.

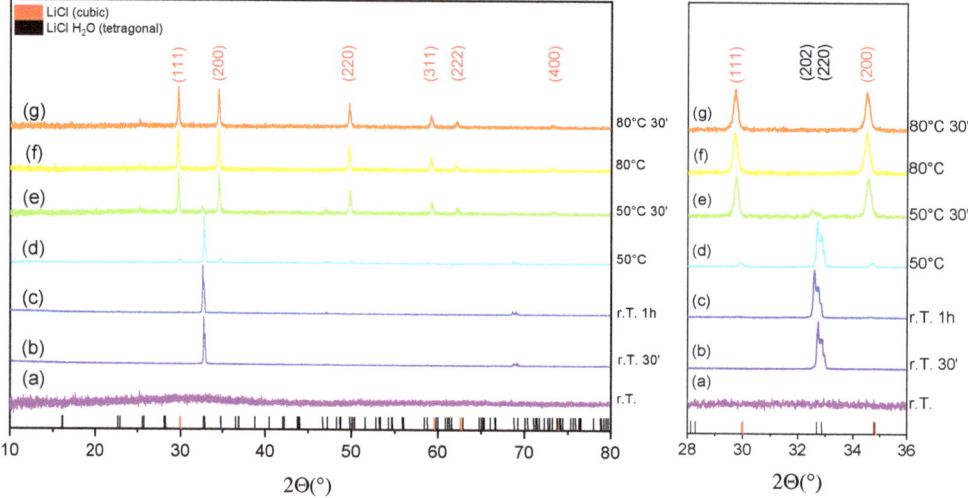

Figure 3. In situ XRD analysis of LiCl-PDMS foam under dry inert atmosphere while heating with
measurement temperature as indicated (°C).

After 30 min under a dry N2 flow, lithium chloride was monohydrated (PDF 04-013-8884) and exhibited a tetragonal structure (P42/nmc; Figure 3 (pattern [b])). After 1 h under the same conditions, the material began to dehydrate (Figure 3 (pattern [c])) and the cubic (Fm-3m) anhydrous phase (PDF 00-004-0664) was present in ~3.7% of the material fraction, as calculated from the Rietveld refinement. As the temperature increased to 50 °C, the most intense reflections indicative of the cubic structure (indexed (111) and (200)) began to be more clearly distinguishable (Figure 3 (pattern [d])), while the ones indicative of the tetragonal structure (indexed (202) and (220)) significantly decreased in intensity. From the phase fraction evolution in Table 1, it is clear that most of the dehydration reaction occurs during the isothermal holding step at 50 °C (Figure 3 (pattern [e])). Indeed, after reaching this temperature, the fraction of the anhydrous phase increased by about 60%. At 80 °C, the material was fully dehydrated (Figure 3 (pattern [f])), and a slight increase in the lattice parameters was observed after the isothermal hold at 80 °C for 30 min (Figure 3 (pattern [g]), Table 1). As lithium chloride transformed from monohydrate to anhydrous, it underwent a significant volume contraction, reduced by ~70% of its initial volume.

Table 1. In situ XRD pattern code, experimental conditions, refined lattice parameters, volume, and phase fraction.

Pattern Code	Experimental Conditions	LiCl·H$_2$O Phase (Tetragonal $P4_2/nmc$)				LiCl Phase (Cubic $Fm\text{-}3m$)		
		Lattice Parameters		Volume (Å^3)	Fraction (%)	Lattice Parameters	Volume (Å^3)	Fraction (%)
		a/$\sqrt{2}$ (Å)	c/$\sqrt{2}$ (Å)			a (=b = c) (Å)		
a	r.T., exposed air	-	-	-	-	-	-	-
b	r.T., N$_2$ flow	5.422 (3)	5.474 (4)	455.33 (6)	100	-	-	0
c	r.T. 30′, N$_2$ flow	5.417 (5)	5.465 (9)	453.68 (8)	96.3	5.106 (3)	133.15 (1)	3.7
d	50 °C, N$_2$ flow	5.401 (6)	5.458 (2)	450.44 (8)	80.8	5.136 (5)	135.48 (4)	19.2
e	50 °C 30′, N$_2$ flow	5.374 (1)	5.442 (6)	444.57 (7)	21.7	5.140 (6)	135.79 (7)	78.3
f	80 °C, N$_2$ flow	-	-	-	-	5.144 (8)	136.11 (4)	100
g	80 °C 30′, N$_2$ flow	-	-	-	-	5.161 (1)	137.46 (8)	100
h	r.T., N$_2$ flow, after cooling	-	-	-	-	5.157 (3)	137.14 (8)	100
i	r.T. 30′, N$_2$ flow, after cooling	-	-	-	-	5.155 (2)	136.99 (9)	100
j	r.T. 1 h exposed air, after cooling	-	-	-	-	-	-	-

After the complete dehydration reaction, the temperature in the XRD chamber was decreased to r.T. and the collected XRD pattern is reported in Figure 4. The analysis after cooling at r.T. (Figure 4, patterns (h) and (i)) showed the persistence of the anhydrous phase obviously due to the lack of humidity in the chamber. In order to rehydrate the material, the LiCl-PDMS foam was exposed to air at r.T. for 1 h. The collected XRD pattern (Figure 4, pattern (j)) showed a completely amorphous structure due to the overhydration of the LiCl salt. This, as expected, is indicative of the complete reaction reversibility.

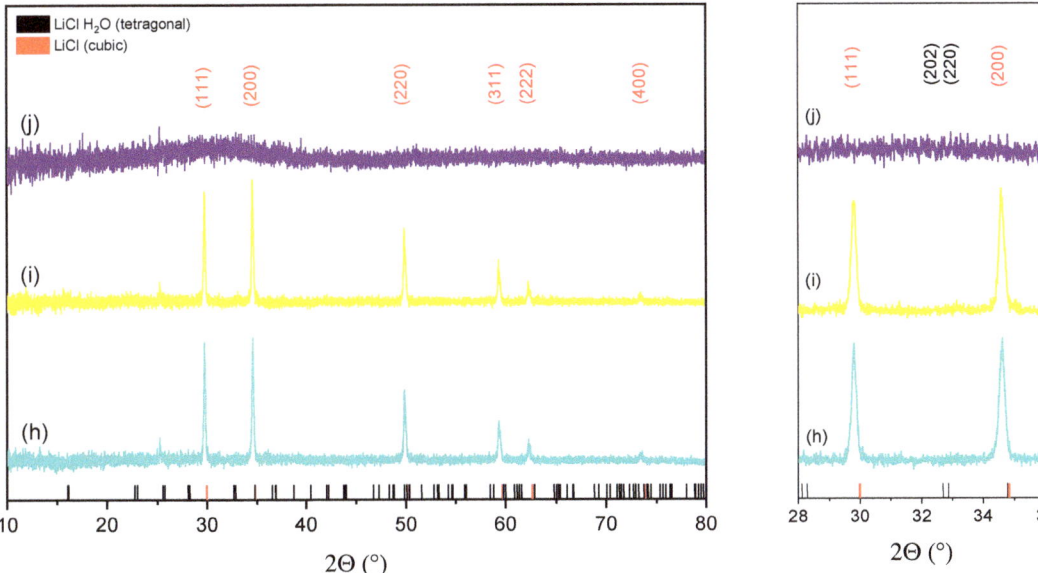

Figure 4. In situ XRD analysis of LiCl-PDMS foam under dry inert atmosphere after cooling at r.T.

3.3. Hydration/Dehydration Thermochemical Behavior

The LiCl-PDMS foam thermochemical behavior during a dehydration/hydration cycle was characterized through a thermogravimetric dynamic vapor system under isothermal conditions at 30 °C by varying the chamber RH in the range of 0–90% and the results are shown in Figure 5. The mass change was normalized to the salt hydrate content (40%) in the foam.

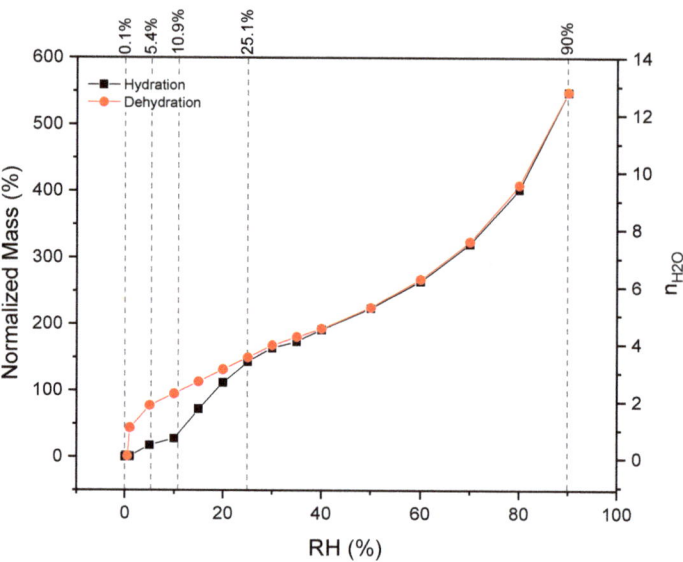

Figure 5. Water vapor hydration/dehydration isotherms (DVS) of the LiCl composite foams at 30 °C.

While increasing the RH in the chamber, the hydration conditions became more favorable for the reaction to proceed and the mass uptake progressively increased. More specifically, the material exhibited a small water uptake at low partial pressure (below 1% RH). At intermediate RH values, between 5 and 25%, a quasi-linear water uptake from ~17 to ~143 wt.% was observed. Finally, above an RH of 25%, a notable mass uptake from ~143 up to 550 wt.% was observed. Taking into account the hydration reaction of LiCl salt, which is

$$\text{LiCl}_{(s)} + \text{H}_2\text{O}_{(g)} \rightleftarrows \text{LiCl} \cdot \text{H}_2\text{O}_{(s)}$$

and the molecular mass of both reagents and products (M_{LiCl} = 42.39 g/mol, $M_{LiCl \cdot H2O}$ = 60.41 g/mol), a theoretical mass uptake of 42.39% is expected when LiCl converts into its monohydrated phase. Hence, it is evident from the gravimetric profile that the full conversion of LiCl into its monohydrated phase occurs at an RH of between 10 and 15%. Above this value, the material is overhydrated and it reaches up to ~13 water molecules. The dehydrated LiCl can exist in the solid form at a low temperature below 20 °C. Hence, the water intake over the monohydrated threshold at the experimental temperature of 30 °C could be caused by the capillary condensation mechanism and the deliquescence phenomenon is expected to take place.

A hysteresis was observable between the dehydration and hydration reactions in the RH range of 0–25%, while above this RH value, the hydration and dehydration profiles almost perfectly overlapped. In general, the presence of hysteresis between the sorption and desorption isotherms indicates that the water diffusion through the material structure is slower as the lattice re-arranges upon hydration [29]. This can be applied to LiCl salt, indeed, as previously observed from in situ XRD analysis (Figures 3 and 4), wherein the material transformed from a completely amorphous structure to a more ordered one when it converts progressively from the (over)hydrated to the mono as well as dehydrated phase.

In Figure 5, we report the RH values adopted for the ESEM conditions that will be discussed in the following Section 3.3.

3.4. Morphological Characterization of LiCl-PDMS Foam during Hydration Reaction

An environmental SEM analysis was conducted on the LiCl-PDMS foam simulating the hydration/dehydration reaction conditions. This approach allowed for visually evaluating the morphological modification while hydrating and dehydrating the LiCl salt embedded in the porous silicon matrix. The dehydration and hydration conditions were obtained with varying temperature and water vapor pressure in the ranges of 5–40 °C and 10–800 Pa, respectively, in order to vary the chamber RH at which the material was exposed.

Initially, the material was fully dehydrated and the RH during the analysis was increased from 0.1% to 91.3% in order to favor the hydration reaction. Afterwards, the RH was decreased to 0.1% to favor the LiCl salt dehydration reaction. Hence, the experiment reproduced the operating conditions of a whole hydration/dehydration cycle. In Figure 6, we report the micrographs at the selected pressure and temperature conditions in the climatic chamber of the microscope. The area selected for the analysis is an inner part of the foam and is indicative of the overall morphology.

At an RH of 0.1%, according to the DVS analysis, the LiCl salt in the PDMS foam should be in its anhydrous phase, and from SEM micrographs on the foam surface, some asperities (1) and bubbles (2) are visible (Figure 6a). It is likely that the asperities consist of loose salt agglomerates, while the bubbles, with a smooth surface and different sizes, might embed the salt in the foam matrix, thus functioning as a covering layer.

By increasing the RH to 5.4%, no visible morphological modifications could be detected (Figure 6b). Nevertheless, at such an RH value, the beginning of the hydration was expected from the hydration profile, with a water mass uptake of ~17 wt.%. As the RH increased to 10.9%, the hydration reaction was more promoted and the bubbles began to swell (Figure 6c), and their volume increased as the RH increased clearly due to the progressive acquisition of more water molecules (Figure 6d,e), in agreement with the DVS analysis.

At an RH of 25.1%, the conversion of LiCl into its monohydrate phase was complete (Figure 5), and above this RH value, the deliquescence phenomenon was expected to occur. The bubbles doubled and tripled their volume at an RH of 91.3% (as indicated by point 3 in Figure 6e). The asperities (red circles in Figure 6e) increased their volume as well as the RH increased, assuming a roundish shape typical of a drop, thus likely indicative of the deliquescence phenomenon. No detachment of the salt from the foam was observable because of the volume expansion. Nevertheless, the loose salt (asperities) appeared to be slightly agglomerated. Additionally, no damages were evident on the PDMS foam where the salt was embedded (bubbles), thus indicating the ability of the material to easily expand and contract as a consequence of the hydration/dehydration cycles and to both effectively protect and retain the embedded salt hydrate.

As final step of the ESEM analysis, the environmental chamber conditions were restored to the initial ones, that is, an RH of 0.1%. Under these conditions, the dehydration reaction was induced and the hydration/dehydration cycle was completed. It could be clearly observed that the material morphology was almost unmodified, that is, no damages due to the retained water and volume expansion/contraction were present, thus preserving the reusability of the foam. The presence of incomplete dehydrated salts cannot be exclued. Remarkably, no defects or cracks were detected, confirming the effectiveness of this approach for obtaining a durable, flexible, and porous-sorbents composite.

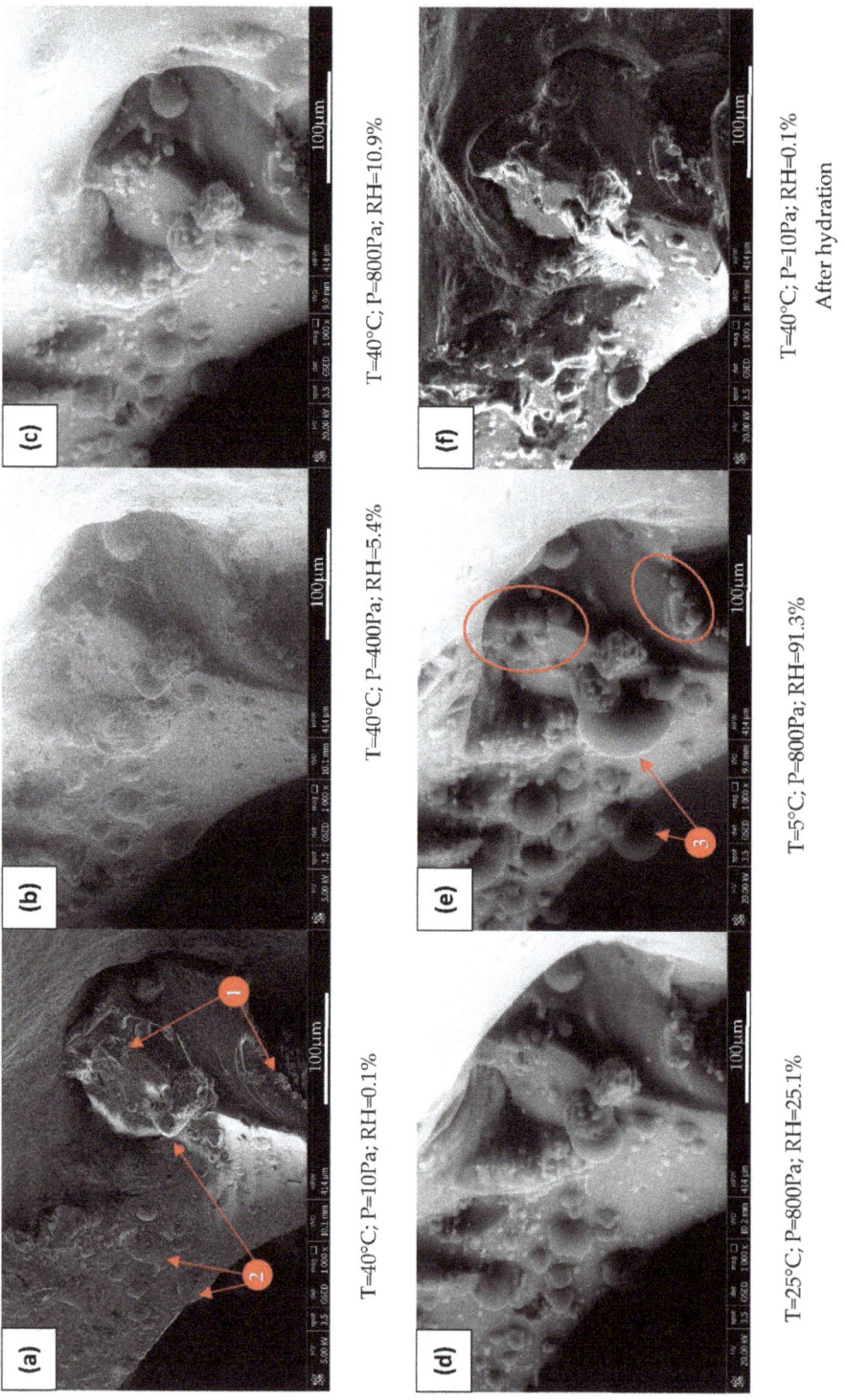

Figure 6. Environmental SEM analysis on LiCl-PDMS foam under varying temperature and water vapor pressure in the ranges of 5–40 °C and 10–800 Pa, respectively.

3.5. Energy Storage/Release Density

To estimate the foam energy storage/release density, the heat release while hydrating was measured with a coupled TG/DSC analysis under pure water vapor atmosphere to simulate the typical operation of a TES system working with a closed cycle. The TG/DSC profile of LiCl-PDMS foam under the temperature drop of 80–35 °C at 12 mbar of water vapor pressure (10 °C of evaporation temperature) is shown in Figure 7a. The material began to uptake water at ~60 °C, reaching a plateau at 35 °C, that is, under 12 mbar of water vapor pressure, it corresponded to an RH of ~22%. Under this condition, the total percentage of the normalized mass gain was ~170%, corresponding to ~4 water molecules, which is in good agreement with the DVS analysis (see Figure 5). Hence, the material was overhydrated. From the integration of the heat flow profile, the LiCl-PDMS foam energy density associated with this water uptake was estimated to be ~1854 kJ/kg$_{foam}$ (or ~323 kWh/m^3). From the deconvolution of the heat flow signal, as shown in Figure 7b, it is evident that four thermal events (exothermic) occurred while the material hydrated. The mass uptake related to the first peak (~40%) is very close to the one occurring when LiCl converted into its monohydrated phase (42.39%), while the second one began at a temperature of 45 °C, that is, under 12 mbar of water vapor pressure, it corresponded to an RH of 12.5%, very close to the value at which a saturated solution of lithium chloride will form an equilibrium (12.4%). Similarly to the DVS analysis, the remaining two peaks could be associated with the water intake over the monohydrated threshold, which can likely be caused by the capillary condensation mechanism in the foam.

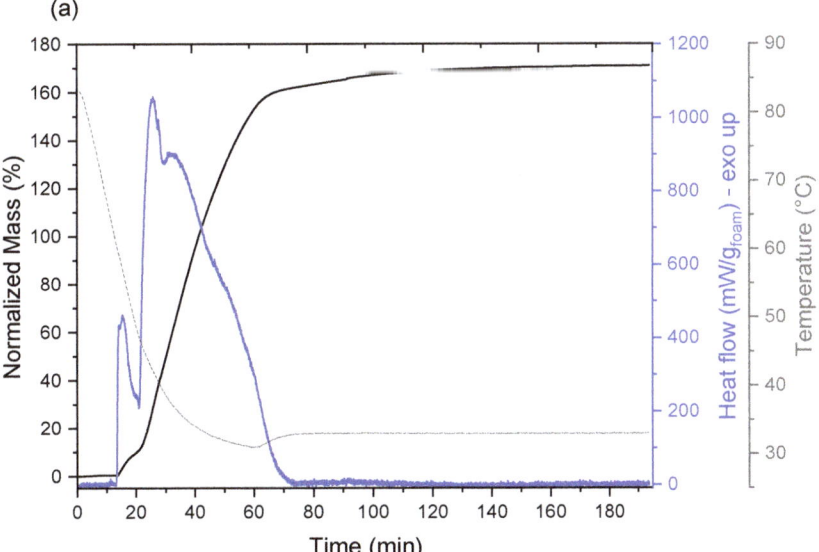

Figure 7. Cont.

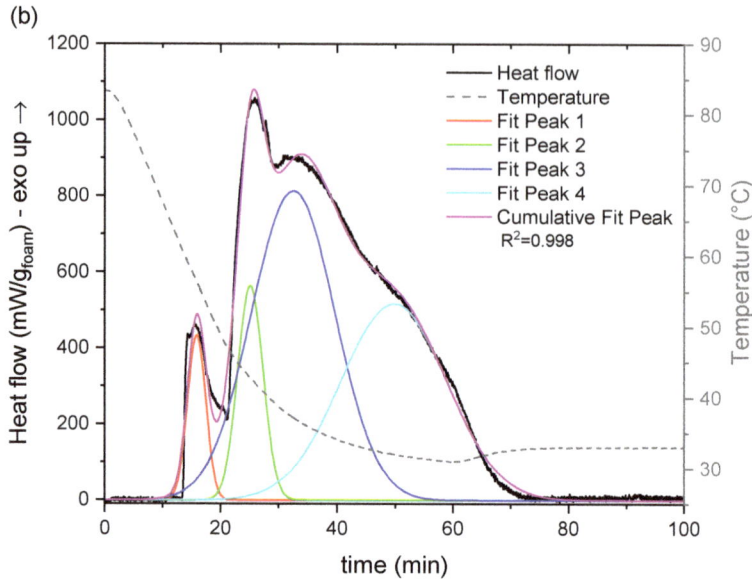

Figure 7. (a) TG/DSC profile of LiCl-PDMS foam under the temperature drop of 80–35 °C at 12 mbar of water vapor pressure, corresponding to 10 °C of evaporation temperature. (b) Deconvolution of DSC profile.

4. Final Remarks

From the morphological analysis, no defects or cracks were detected, confirming the effectiveness of this approach for obtaining a durable, flexible, and porous-sorbents composite. Nevertheless, it has to be pointed out that LiCl is not completely embedded by the PDMS foam but, rather, it is partially deposited on the external foam surface. Critically, this superficial salt could affect the overall TES performances due to possible corrosion processes. Hence, this issue should be addressed. Additionally, the material stability at long aging cycles is a key point for thermochemical energy storage applications and will be addressed in the following studies.

With respect to some of the most recently investigated LiCl-based composites (see Table 2), the LiCl-PDMS foam exhibits larger heat storage/release capacity and energy density under both similar testing conditions and similar LiCl content. Hence, it can be considered a competitive candidate among other LiCl salt hydrate composites.

Table 2. Comparison of some LiCl-based composites' energy densities and testing conditions.

Li Composite	LiCl Content (wt.%)	Stored/Released Heat (kJ/kg$_{composite}$)	Energy Storage Density (kWh/m^3)	Charging/Discharging Conditions			Ref.
				T_{char} (°C)	T_{dis} (°C)	RH_{dis} (%)	
Li-PDMS foam	40	1854	323	80	35	22	this work
LiCl/activated alumina	14.68	874	318	120	20	80	[18]
LiCl/Siogel	31.10	1000	166	66	35	15	[30]
LiCl/Vermiculite	45.20	1250	111	66	35	15	[30]
LiCl/MWCNTs	41–42	1700	n.a.	75–46	65–35	15	[20]

5. Conclusions

Here, we investigated a composite material based on silicone vapor-permeable foam filled with 40 wt.% of LiCl (LiCl-PDMS), aiming at confining the salt in a matrix to prevent deliquescence-related issues but without inhibiting the vapour flow. Specifically, we studied the structural and morphological modification of the composite foam during the hydration/dehydration cycles through in situ XRD and ESEM analysis. In situ XRD enabled the material dehydration reaction from room temperature up to 80 °C, while the environmental SEM analysis simulated the hydration/dehydration reaction conditions, thus allowing for the evaluation of the interaction between the salt and the water vapor environment in the confined silicon matrix. The results show an effective embedding of the material, which limits the salt solution release when overhydrated. Additionally, the flexibility of the matrix allows for the volume shrinkage/expansion of the salt caused by the cyclic dehydration/hydration reactions without any damages to the foam structure. Indeed, as lithium chloride transformed from monohydrate to anhydrous, a significant volume contraction was observable (~70% of its initial volume). The LiCl-PDMS foam energy density was measured with a customized coupled thermogravimetric/differential scanning calorimetric analysis (TG/DSC) during the hydration phase and the estimated energy density value was ~1854 kJ/kg (or ~323 kWh/m^3), thus making it a competitive candidate among other LiCl salt hydrate composites.

Author Contributions: Conceptualization, L.C. and E.P.; investigation, E.P., A.F., D.P. and E.M.; data curation, E.M. and D.P.; validation, L.C., E.P., A.F. and E.M.; writing—original draft preparation, E.M.; writing—review and editing, E.M., E.P. and L.C. All authors have read and agreed to the published version of the manuscript.

Funding: This research study was funded by the Italian Ministry of University and Research (MUR), program PON R&I 2014/2020—Avviso n. 1735 del 13 July 2017—PNR 2015/2020, under project "NAUSICA—NAvi efficienti tramite l'Utilizzo di Soluzioni tecnologiche Innovative e low CArbon", CUP: B45F21000680005.

Conflicts of Interest: The authors declare no conflict of interest. The funders had no role in the design of the study; in the collection, analyses, or interpretation of data; in the writing of the manuscript, or in the decision to publish the results.

References

1. Gielen, D.; Boshell, F.; Saygin, D.; Bazilian, M.D.; Wagner, N.; Gorini, R. The role of renewable energy in the global energy transformation. *Energy Strategy Rev.* **2019**, *24*, 38–50. [CrossRef]
2. IRENA. *Global Energy Transformation: A roadmap to 2050, International Renewable Energy Agency*; IRENA: Abu Dhabi, United Arab Emirates, 2018; p. 12.
3. Liu, H.; Wang, W.; Zhang, Y. Performance gap between thermochemical energy storage systems based on salt hydrates and materials. *J. Clean. Prod.* **2021**, *313*, 127908. [CrossRef]
4. Donkers, P.; Sögütoglu, L.; Huinink, H.; Fischer, H.; Adan, O. A review of salt hydrates for seasonal heat storage in domestic applications. *Appl. Energy* **2017**, *199*, 45–68. [CrossRef]
5. Trausel, F.; de Jong, A.-J.; Cuypers, R. A Review on the Properties of Salt Hydrates for Thermochemical Storage. *Energy Procedia* **2014**, *48*, 447–452. [CrossRef]
6. Courbon, E.; D'Ans, P.; Skrylnyk, O.; Frère, M. New prominent lithium bromide-based composites for thermal energy storage. *J. Energy Storage* **2020**, *32*, 101699. [CrossRef]
7. Wang, Q.; Xie, Y.; Ding, B.; Yu, G.; Ye, F.; Xu, C. Structure and hydration state characterizations of MgSO$_4$-zeolite 13x composite materials for long-term thermochemical heat storage. *Sol. Energy Mater. Sol. Cells* **2019**, *200*, 110047. [CrossRef]
8. Casey, S.P.; Elvins, J.; Riffat, S.; Robinson, A. Salt impregnated desiccant matrices for 'open' thermochemical energy storage—Selection, synthesis and characterisation of candidate materials. *Energy Build.* **2014**, *84*, 412–425. [CrossRef]
9. Farcot, L.; Le Pierrès, N.; Fourmigué, J.-F. Experimental investigation of a moving-bed heat storage thermochemical reactor with SrBr2/H2O couple. *J. Energy Storage* **2019**, *26*, 101009. [CrossRef]
10. Liu, H.; Nagano, K.; Togawa, J. A composite material made of mesoporous siliceous shale impregnated with lithium chloride for an open sorption thermal energy storage system. *Sol. Energy* **2015**, *111*, 186–200. [CrossRef]
11. Marín, P.; Milian, Y.; Ushak, S.; Cabeza, L.; Grágeda, M.; Shire, G. Lithium compounds for thermochemical energy storage: A state-of-the-art review and future trends. *Renew. Sustain. Energy Rev.* **2021**, *149*, 111381. [CrossRef]
12. Scapino, L.; Zondag, H.A.; Van Bael, J.; Diriken, J.; Rindt, C.C. Energy density and storage capacity cost comparison of conceptual solid and liquid sorption seasonal heat storage systems for low-temperature space heating. *Renew. Sustain. Energy Rev.* **2017**, *76*, 1314–1331. [CrossRef]
13. Khvorostyanov, V.I.; Curry, J.A. Deliquescence and Efflorescence in Atmospheric Aerosols. In *Thermodynamics, Kinetics, and Microphysics of Clouds*; Cambridge University Press: Cambridge, UK, 2014; pp. 547–576, ISBN 9781139060004.
14. N'Tsoukpoe, K.E.; Schmidt, T.; Rammelberg, H.U.; Watts, B.A.; Ruck, W.K. A systematic multi-step screening of numerous salt hydrates for low temperature thermochemical energy storage. *Appl. Energy* **2014**, *124*, 1–16. [CrossRef]

15. Kohler, T.; Biedermann, T.; Müller, K. Experimental Study of $MgCl_2$ 6 H_2O as Thermochemical Energy Storage Material. *Energy Technol.* **2018**, *6*, 1935–1940. [CrossRef]
16. Druske, M.-M.; Fopah-Lele, A.; Korhammer, K.; Rammelberg, H.U.; Wegscheider, N.; Ruck, W.; Schmidt, T. Developed Materials for Thermal Energy Storage: Synthesis and Characterization. *Energy Procedia* **2014**, *61*, 96–99. [CrossRef]
17. Zbair, M.; Bennici, S. Survey Summary on Salts Hydrates and Composites Used in Thermochemical Sorption Heat Storage: A Review. *Energies* **2021**, *14*, 3105. [CrossRef]
18. Zhang, Y.; Wang, R.; Li, T. Thermochemical characterizations of high-stable activated alumina/LiCl composites with multistage sorption process for thermal storage. *Energy* **2018**, *156*, 240–249. [CrossRef]
19. Grekova, A.; Gordeeva, L.; Aristov, Y. Composite sorbents "Li/Ca halogenides inside Multi-wall Carbon Nano-tubes" for Thermal Energy Storage. *Sol. Energy Mater. Sol. Cells* **2016**, *155*, 176–183. [CrossRef]
20. Grekova, A.D.; Gordeeva, L.G.; Lu, Z.; Wang, R.; Aristov, Y.I. Composite "LiCl/MWCNT" as advanced water sorbent for thermal energy storage: Sorption dynamics. *Sol. Energy Mater. Sol. Cells* **2018**, *176*, 273–279. [CrossRef]
21. Yu, N.; Wang, R.; Lu, Z.; Wang, L. Development and characterization of silica gel–LiCl composite sorbents for thermal energy storage. *Chem. Eng. Sci.* **2014**, *111*, 73–84. [CrossRef]
22. Frazzica, A.; Brancato, V.; Caprì, A.; Cannilla, C.; Gordeeva, L.; Aristov, Y. Development of "salt in porous matrix" composites based on LiCl for sorption thermal energy storage. *Energy* **2020**, *208*, 118338. [CrossRef]
23. Piperopoulos, E.; Calabrese, L.; Bruzzaniti, P.; Brancato, V.; Palomba, V.; Caprì, A.; Frazzica, A.; Cabeza, L.F.; Proverbio, E.; Milone, C. Morphological and Structural Evaluation of Hydration/Dehydration Stages of $MgSO_4$ Filled Composite Silicone Foam for Thermal Energy Storage Applications. *Appl. Sci.* **2020**, *10*, 453. [CrossRef]
24. Brancato, V.; Calabrese, L.; Palomba, V.; Frazzica, A.; Fullana-Puig, M.; Solé, A.; Cabeza, L.F. $MgSO_4 \cdot 7H_2O$ filled macro cellular foams: An innovative composite sorbent for thermo-chemical energy storage applications for solar buildings. *Sol. Energy* **2018**, *173*, 1278–1286. [CrossRef]
25. Calabrese, L.; Brancato, V.; Palomba, V.; Frazzica, A.; Cabeza, L.F. Magnesium sulphate-silicone foam composites for thermochemical energy storage: Assessment of dehydration behaviour and mechanical stability. *Sol. Energy Mater. Sol. Cells* **2019**, *200*, 109992. [CrossRef]
26. Toby, B.H.; Von Dreele, R.B. GSAS-II: The genesis of a modern open-source all purpose crystallography software package. *J. Appl. Crystallogr.* **2013**, *46*, 544–549. [CrossRef]
27. Frazzica, A.; Sapienza, A.; Freni, A. Novel experimental methodology for the characterization of thermodynamic performance of advanced working pairs for adsorptive heat transformers. *Appl. Therm. Eng.* **2014**, *72*, 229–236. [CrossRef]
28. Brancato, V.; Gordeeva, L.G.; Sapienza, A.; Palomba, V.; Vasta, S.; Grekova, A.D.; Frazzica, A.; Aristov, Y.I. Experimental characterization of the LiCl/vermiculite composite for sorption heat storage applications. *Int. J. Refrig.* **2019**, *105*, 92–100. [CrossRef]
29. Tieger, E.; Kiss, V.; Pokol, G.; Finta, Z.; Dušek, M.; Rohlíček, J.; Skořepová, E.; Brázda, P. Studies on the crystal structure and arrangement of water in sitagliptinl-tartrate hydrates. *CrystEngComm* **2016**, *18*, 3819–3831. [CrossRef]
30. Brancato, V.; Gordeeva, L.; Caprì, A.; Grekova, A.; Frazzica, A. Experimental Comparison of Innovative Composite Sorbents for Space Heating and Domestic Hot Water Storage. *Crystals* **2021**, *11*, 476. [CrossRef]

Article

Thermal Stability of Ionic Liquids: Effect of Metals

Francesca Nardelli [1,2], Emilia Bramanti [2], Alessandro Lavacchi [3], Silvia Pizzanelli [2,4], Beatrice Campanella [2], Claudia Forte [2,4], Enrico Berretti [3] and Angelo Freni [2,*]

[1] Dipartimento di Chimica e Chimica Industriale, Università di Pisa, Via G. Moruzzi 13, 56124 Pisa, Italy; francesca.nardelli@dcci.unipi.it
[2] Istituto di Chimica dei Composti OrganoMetallici, Consiglio Nazionale delle Ricerche, Via G. Moruzzi 1, 56124 Pisa, Italy; emilia.bramanti@pi.iccom.cnr.it (E.B.); silvia.pizzanelli@pi.iccom.cnr.it (S.P.); beatrice.campanella@pi.iccom.cnr.it (B.C.); claudia.forte@pi.iccom.cnr.it (C.F.)
[3] Istituto di Chimica dei Composti OrganoMetallici, Consiglio Nazionale delle Ricerche, Via Madonna del Piano 10, 50019 Firenze, Italy; alessandro.lavacchi@iccom.cnr.it (A.L.); enrico.berretti@iccom.cnr.it (E.B.)
[4] Centre for Instrument Sharing (CISUP), University of Pisa, Lungarno Pacinotti 43, 56126 Pisa, Italy
* Correspondence: angelo.freni@pi.iccom.cnr.it

Abstract: We investigated the thermal stability and corrosion effects of a promising ionic liquid (IL) to be employed as an advanced heat transfer fluid in solar thermal energy applications. Degradation tests were performed on IL samples kept in contact with various metals (steel, copper and brass) at 200 °C for different time lengths. Structural characterization of fresh and aged IL samples was carried out by high-resolution magic angle spinning nuclear magnetic resonance and Fourier transform infrared spectroscopic analyses, while headspace gas chromatography–mass spectrometry was employed to evaluate the release of volatile organic compounds. The combination of the above-mentioned techniques effectively allowed the occurrence of degradation processes due to aging to be verified.

Keywords: ionic liquids; heat storage; thermal stability; HRMAS NMR; FTIR

1. Introduction

Ionic liquids (ILs) are a group of compounds that are attracting increasing interest in many fields of application, thanks to the possibility of combining different anions and cations, thus allowing the design of new materials with optimal chemical–physical properties for specific applications, especially in the energy sector [1,2]. In particular, ILs are suggested as promising working fluids in solar energy technologies, thanks to their high heat capacity, low melting point and relatively high density in the typical operating conditions of solar thermal energy systems [3–6]. Further attractive features of ILs are the high chemical stability, non-flammability, and the low impact on the environment and on health; this feature derives from their negligible vapor pressure, which limits their release in the atmosphere [7]. Given the wide application potential, evaluation of ILs' thermal stability is fundamental for their implementation in solar energy systems as working fluids [8–10].

Most of the thermal stability studies available in the literature are based on dynamic thermogravimetric (TG) analyses [11,12]. However, several experimental parameters, such as sample mass, pre-treatment conditions, heating rate and testing atmosphere (inert gas or open air), can affect measurement consistency [13]; therefore, TG analysis appears to be more appropriate for comparative thermal stability studies [14], and certainly cannot provide a deep insight into the modifications of the IL structure due to thermal stress. Another issue of relevance is metal corrosion in the presence of ILs. In fact, several R&D activities in the field have focused on the investigation of the corrosion behavior of different metals in contact with ILs, and on the evaluation of the release of volatile compounds during operation in solar thermal devices and processes [15–19].

To address the previously mentioned issues, in this paper, we present a multi-technique approach to identify possible degradation products of a promising IL subjected to thermal aging in the absence or presence of different metals. Specifically, we have characterized the ionic liquid N-tributyl-N-methylammonium bis(trifluoromethanesulfonyl)imide ([TBuMA][NTF$_2$]) after thermal treatment at T = 200 °C for 4, 24 and 168 h in contact with AISI 304 steel, copper or brass, as well as in the absence of metals. The temperature used for aging was selected because it is the standard operating temperature of common diathermic oils used as heat transfer fluids. This specific ammonium-based IL compound was chosen for its potential application as a heat transfer fluid [5,12], and for the stability of the anion [20], which has been reported to withstand temperatures up to 400 °C by thermogravimetry. However, to the best of our knowledge, there is no indication of the thermal stability of NTF$_2$ coupled with a quaternary ammonium salt. The characterization of the degraded IL was carried out by high-resolution magic angle spinning nuclear magnetic resonance (HRMAS NMR) and Fourier transform infrared (FTIR) spectroscopies, while headspace gas chromatography–mass spectrometry (HS-GC-MS) was employed to estimate the concentration of volatile compounds produced. HRMAS NMR is suitable for the characterization of highly viscous liquids. In this tecnique, the use of magic angle spinning allows highly resolved spectra to be obtained, which is not feasible using standard solution NMR spectroscopy, due to the presence of residual interactions and magnetic susceptibility issues [21].

2. Materials and Methods

2.1. Materials

The ionic liquid N-tributyl-N-methylammonium bis(trifluoromethanesulfonyl)imide ($C_{15}H_{30}F_6N_2O_4S_2$), CAS number 405514-94-5; MW 480.53, Figure 1a, was purchased from Solvionic (Am3408a). Purity of the IL was 99.9%.

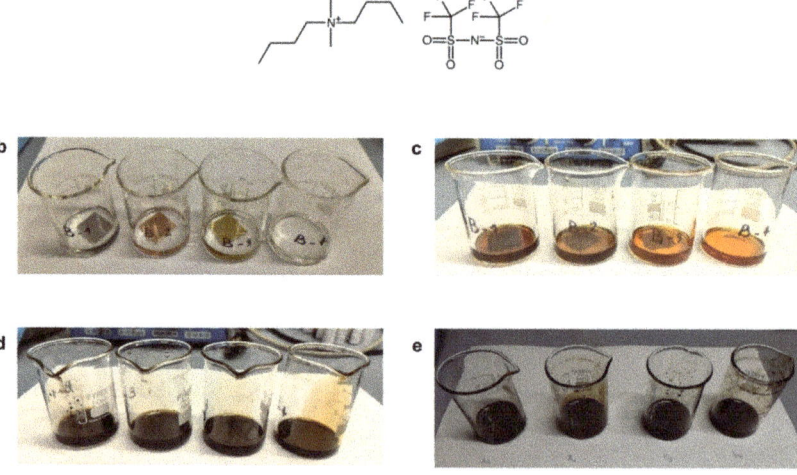

Figure 1. (a) Structure of the ionic liquid [TBuMA][NTF$_2$]. (b) In sequence, from left to right, samples of the ionic liquid in the presence of steel, copper and brass metal plates and with no metal plate. The same samples were heated at 200 °C for 4 h (c), 24 h (d) and 168 h (e).

The degradation procedure was performed as described in the following. Six milliliters of [TBuMA][NTF$_2$] was heated in an oven at 200 °C for 7 days with or without a steel, copper or brass metal plate (2 × 2 cm). At selected times (4, 24 and 168 h; Figure 1b–e, respectively), 1 mL of IL was sampled for the analyses (FTIR, HS-GC-MS and HRMAS NMR spectroscopy).

Table 1 summarizes the thermally treated samples analyzed and the code used throughout the text. The code of the initial non-heated sample is B.

Table 1. Sample codes of the IL samples.

Metal	4 h at 200 °C	24 h at 200 °C	168 h at 200 °C
1 (steel)	1B4	1B24	1B168
2 (copper)	2B4	2B24	2B168
3 (brass)	3B4	3B24	3B168
4 (no metal)	4B4	4B24	4B168

2.2. FTIR Spectroscopy

Infrared spectra were recorded in reflectance mode by using a Perkin–Elmer Frontiers FTIR Spectrophotometer, equipped with a universal attenuated total reflectance (ATR) accessory and a triglycine sulphate TGS detector. Three replicates (3–5 µL of IL for each measurement) were performed after background acquisition. For each sample, 32 scans were recorded, averaged and Fourier transformed to produce a spectrum with a nominal resolution of 4 cm^{-1}.

2.3. HS-GC-MS Analysis

HS-GC-MS analyses were performed using an Agilent 6850 gas chromatograph, equipped with a split/splitless injector, in combination with an Agilent 5975c mass spectrometer. A CTC CombiPAL autosampler was employed for HS sampling. Vials with 1 g of sample were incubated at 80 °C for 15 min. A 0.5 mL HS volume was then sampled (gas-tight syringe held at 85 °C) and injected into the GC. The syringe was then flushed with helium. The inlet liner (internal diameter of 1 mm) was held at 200 °C and the injection was performed in splitless mode. Compounds were separated on a polar column (DB-WAX ultra-inert; length: 30 m; stationary phase: bonded polyethylene glycol; 0.25 mm inner diameter; 0.50 µm coating) using the following temperature program: 10 min at 30 °C, then increased by 5 °C/min to 60 °C (held for 2 min) followed by an increase of 10 °C/min to 240 °C (held for 9 min). The temperature of the transfer line was set at 250 °C. After GC separation, compounds were ionized in positive EI, and the acquisition was performed in full scan mode. Spectral identification was performed when the spectra and the NIST spectral mass library (NIST 05) combined with our in-house library matched with a spectral similarity >90%. Results are reported as relative intensity (counts).

2.4. HRMAS NMR Spectroscopy

NMR spectra were acquired on a Bruker AVANCE NEO NMR Spectrometer, working at a ^{1}H Larmor frequency of 500.13 and 125.77 MHz for ^{1}H and ^{13}C nuclei, respectively, and using an HRMAS probe. All samples were spun at 6 kHz. The samples were dissolved in DMSO-d$_6$ (99.7% deuterated, Sigma) (1:1 volume ratio) to provide the lock signal and to reduce their viscosity, thus facilitating their insertion in the rotors; TMS was added to each mixture for ^{1}H spectral referencing. Following this, 50 µL of each mixture was transferred to an HRMAS rotor for NMR analysis. ^{1}H spectra were acquired on all samples using a relaxation delay of 1 s and several scans ranging from 128 to 1000 depending on the sample. ^{13}C spectra were acquired on samples B, 1B168, 2B168 and 4B168 using the Bruker *zg30pg* pulse sequence for NOE enhancement of carbon nuclei signals. A relaxation delay of 2 s was used and 4k scans were accumulated. One-dimensional (1D) selective ^{1}H total correlation spectroscopy (TOCSY) and two-dimensional (2D) ^{1}H-^{13}C heteronuclear single quantum coherence (HSQC) experiments were also performed on sample 4B168. For the TOCSY experiments, the Bruker *seldigpzs* pulse sequence was used, with a Gaussian shaped 180° pulse (Bruker pulse shape: Gaus1_180r.1000) for selective excitation, a relaxation delay of 1 s, and a mixing time of 80 ms. ^{1}H–^{13}C HSQC were obtained by employing the Bruker *hsqcetgpsisp2.2* pulse sequence, with a relaxation delay of 1 s. For all experiments, a ^{1}H 90° pulse of 7 µs and a ^{13}C 90° pulse of 12 µs were used. All experiments were performed at 298 K.

3. Results and Discussion

Figure 1 shows that, after 4 h, all the samples displayed a brown color, indicating that the thermal treatment degrades [TBuMA][NTF$_2$]. The color was more intense in the presence of metals, particularly steel and copper, and became darker with longer heating times. To understand the decomposition pathway(s), FTIR and HRMAS NMR experiments were performed on all the samples. Figure 2 shows the ATR-FTIR spectra of [TBuMA][NTF$_2$] after 4, 24 and 168 h (samples 4B4, 4B24 and 4B168) of thermal treatment without metal plates. The spectrum of untreated IL (sample B) is also reported for comparison.

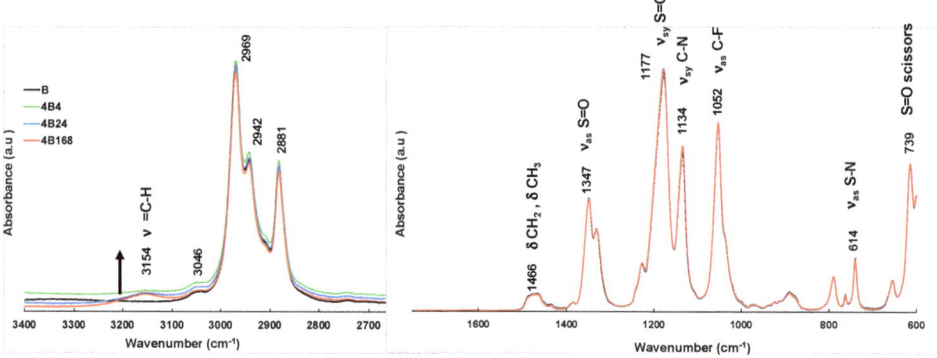

Figure 2. Representative ATR-FTIR spectra of B, 4B4, 4B24 and 4B168 samples in the 3400–2650 and 1740–600 cm^{-1} regions.

As far as the IL anion is concerned, the strong absorptions at 1347 and 1177 cm^{-1} are attributed to asymmetric and symmetric S=O stretching vibrations, respectively, the band at 1052 cm^{-1} to asymmetric C–F stretching, the band at 739 cm^{-1} to asymmetric S–N stretching, and the band at 614 cm^{-1} to S=O scissoring. Specific TBuMA cation signals are expected at 1134 cm^{-1}, ascribable to symmetric C–N stretching, around 1470 cm^{-1} due to methyl and methylene C–H bending, and at 1465 and 1378 cm^{-1} due to C–H scissoring and methyl rocking, respectively. The spectra of all the samples are basically identical, except for slight differences in the 3250–3000 cm^{-1} region (Figure 3), suggesting that the anion is not affected by the thermal treatment, and indicating a major involvement of the cation in the thermal degradation. Inspection of this region highlights that the original structure of the TBuMA cation, characterized by the large band at 3348 cm^{-1} and the shoulder at 3040 cm^{-1}, due to ammonium absorptions, is modified after thermal treatment. The most significant changes are the decrease in the band at 3348 cm^{-1} and the increase in the peak in the region between 3200 and 3100 cm^{-1}, both in the presence and absence of metals. This peak has been assigned to the medium intensity band of unsaturated hydrogen stretches (C=C–H) [22], and suggests the formation of alkenes.

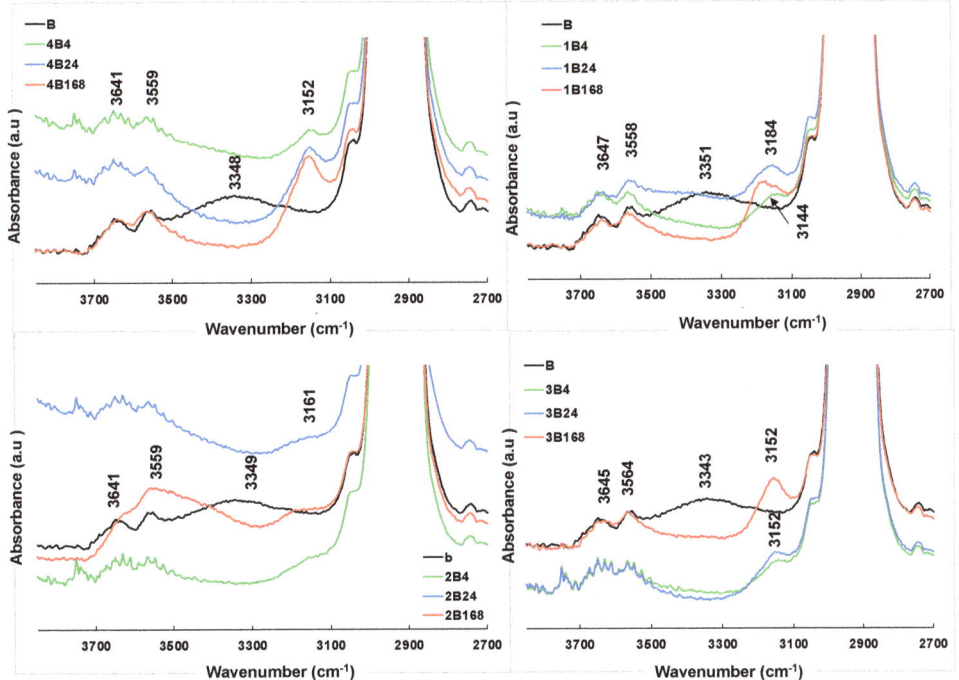

Figure 3. ATR-FTIR spectra of all samples in the 3400–2650 cm^{-1} region.

Figure 4 shows the trend of the area of the band at 3153 cm^{-1} (3208–3103 cm^{-1} baseline points) of the IL spectra with or without metal plates, as a function of the duration of the thermal treatment.

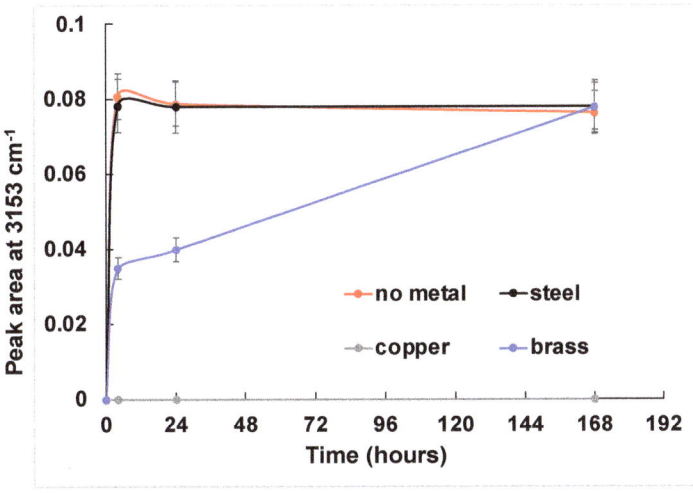

Figure 4. Area of the peak at 3153 cm^{-1} in the IL spectra after 4, 24 and 168 h of thermal treatment with or without metal plates as a function of incubation time.

Further insight on the degradation process was gained from HRMAS NMR spectroscopy. The comparison of the ^{1}H NMR spectrum of the original compound [TBuMA][NTF$_2$] with the spectra recorded on [TBuMA][NTF$_2$] after heating for 4, 24, and 168 h (samples 4B4, 4B24, and 4B168) reveals that the signals of the original cation remain dominant, even at the longest heating time. Low-intensity peaks, due to the degradation products, appear in the samples subjected to thermal treatment, and their intensity tends to increase with the heating time. Figure 5 shows the NMR spectra of samples B (traces a and b) and 4B168 (traces c and d). Four new signals, resonating at 9.16, 8.31, 2.75 and 2.57 ppm, appear in the latter. Additional signals of lower intensity appear in the region between 5 and 6 ppm upon heating. Complete characterization of the degradation compounds was accomplished by the analysis of ^{1}H, 1D selective ^{1}H TOCSY, ^{13}C and 2D ^{1}H–^{13}C HSQC spectra of 4B168 (Figures S1 and S2). The assignments of ^{1}H and ^{13}C NMR signals are reported in Tables S1 and S2.

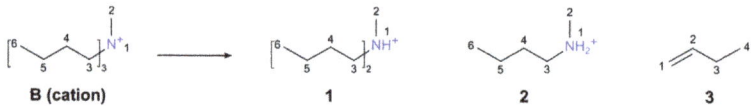

Scheme 1. Decomposition pathways of cation TBuMA. The labeling on each compound is used for the assignment of the NMR signals.

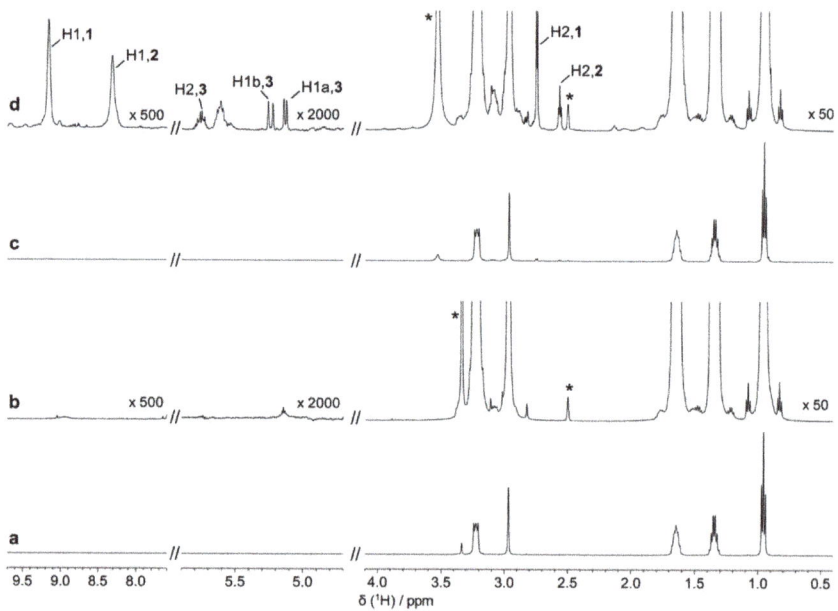

Figure 5. ^{1}H HRMAS NMR spectra of [TBuMA][NTF$_2$] (**a,b**) and 4B168 (**c,d**); in (**b,d**), the vertical scale used in (**a,c**) is expanded by the factors reported on each spectral region. Selected signals of the degradation compounds are labeled as "Hi,j", with i representing the atom number and j indicating the degradation product (Scheme 1). Signals of water protons and residual protons of deuterated DMSO are marked with asterisks.

The dominant thermal degradation products are N-dibutyl-N-methylammonium (**1**) and N-butyl-N-methylammonium (**2**), as outlined in Scheme 1; these compounds are compatible with the Hoffman elimination of one or two alkyl chains from the original cation [23]. For these compounds, the signals due to the hydrogen atoms labeled as H1 and H2 in Scheme 1 are clearly observable. Additional signals, characterized by lower intensities, were assigned to 1-butene (Scheme 1, compound 3), in agreement with the hypothesized degradation pathway. These assignments are indicated in Figure 5, where the signals are labeled as "Hi,j", with i representing the atom number and j the degradation product. The amount of compounds 1 and 2 was determined from the ^{1}H spectra of all the samples, with respect to the amount of the original cation B, using the integrals of the H2,1 and H2,2 signals. Figure 6 shows the values of these integrals as a function of the heating time, where the intensity of the corresponding signal of the non-degraded ionic liquid, occurring at 2.97 ppm, is arbitrarily set to 100. It was, thus, found that, even at the longest heating time, about six molecules of compound 1 and two molecules of compound 2 were present every 100 molecules of ionic liquid cation. This estimate agrees with that obtained using H1,1 and H1,2 signals. Interestingly, the concentration of compound 2 increased at a slower rate than compound 1, in agreement with the fact that the formation of compound 2 requires the preliminary formation of compound 1. Moreover, the amount of compound 3 was always much lower than that expected on the basis of the stoichiometry of the degradation pathways, as clearly evident from the large scaling factor necessary to visualize the signals of this compound (Figure 5d). This can be explained by the volatility of 1-butene or its further degradation/reactions. The presence of 1-butene is compatible with the alkene absorption band between 3200 and 3100 cm^{-1} detected by FTIR (3090 cm^{-1} in the FTIR spectrum of butene in the vapor gas phase). Unfortunately, the absence of specific absorption lines prevented the detection of compounds 1 and 2 in the FTIR spectra.

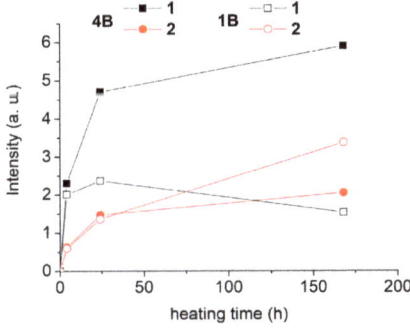

Figure 6. Intensity of H2,1 (black squares) and H2,2 (red circles) signals as a function of the heating time, as obtained from the ^{1}H HRMAS NMR spectra of B, 4B4, 4B24 and 4B168 (filled symbols) and of B, 1B4, 1B24 and 1B168 (empty symbols). The intensity value is relative to that of H2 protons of the non-degraded ionic liquid, which was set to 100 in all spectra.

The ^{1}H NMR spectra recorded after heating for up to 168 h in the presence of metal plates do not show significant differences with respect to the 4B168 spectrum (Figure 7). The main signals of the NMR spectrum of 4B168 are those of the original ionic liquid. For both steel and copper, the spectral lines are broadened, with steel inducing larger line broadening of the signals compared to copper. This broadening, not observed in the case of brass, is probably due to the presence of dissolved paramagnetic metal ions resulting from corrosion of the metal/alloy. Figure 8 shows that the signals of the degradation compounds 1, 2 and 3 are also present in the spectra of samples 1B168 and 3B168, whereas, in the case of 2B168, only sharp signals, due to compound 3, are observed. In the latter case, if compounds 1 and 2 are present, their concentration is below detection limits. However, the occurrence of 1-butene signals suggests that the elimination reactions sketched in Scheme 1 also take place in the presence of copper. The absence of the H1,1 and H1,2 signals could be explained by hypothesizing that the ammonium compounds 1 and 2 release H$^+$ and the amine formed coordinates to a copper ion. A similar mechanism has been suggested to rationalize the extraction of Cu^{2+} ions from aqueous solutions using protic ammonium ionic liquids [24]. However, due to the large linewidth of the signals, it was not possible to confirm this hypothesis.

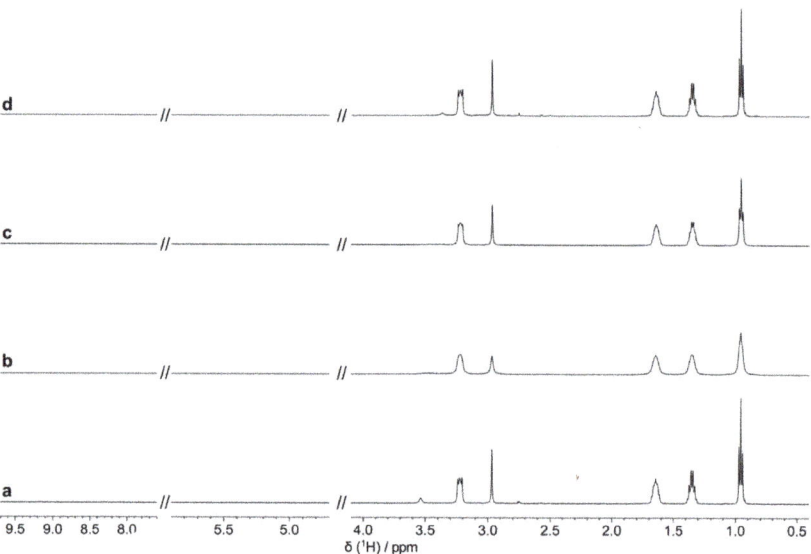

Figure 7. ^{1}H HRMAS NMR spectra of 4B168 (**a**), 1B168 (**b**), 2B168 (**c**), and 3B168 (**d**).

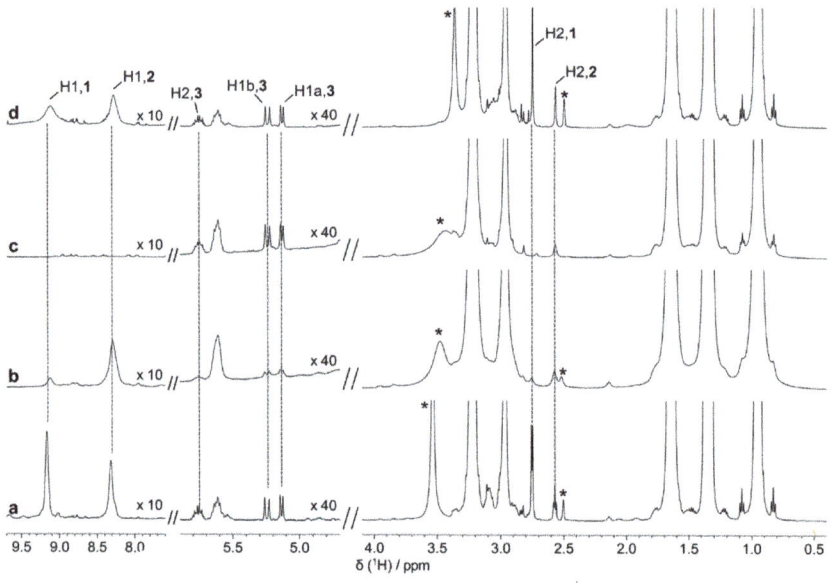

Figure 8. Expansions of ^{1}H HRMAS NMR spectra of 4B168 (**a**), 1B168 (**b**), 2B168 (**c**), and 3B168 (**d**). Expansion factors are reported for each region. Selected signals of the degradation compounds are labeled using the same notation as that applied in Figure 5. Water (3.3–3.6 ppm) and residual DMSO-d_5 (2.50 ppm) signals are marked with asterisks.

For the samples degraded in the presence of steel, the kinetics of formation of compounds 1 and 2 were monitored. Figure 6 shows the trends of the intensity of the H2,1 and

H2,2 signals as a function of heating time in the samples degraded in the presence and absence of metals. The kinetics of formation of compound 2 from compound 1 seems to be accelerated by steel.

HS-GCMS analysis was performed on the 168 h aged samples, with the aim of observing possible volatile degradation compounds in the most extreme condition. Figure 9 shows the peak area of nine main compounds identified in samples B, 4B168, 1B168, 2B168 and 3B168; these are cyclohexane, butanal, ethyl acetate, tert-butanol, methyl-vinyl ketone, N,-N-dimethylformamide, 2-ethyl-1-hexanol and benzothiazole. However, 1-butene was not detected, likely because it was lost in the pre-analytical phase, considering its high volatility, or because of its oxidation. The relevant result is, indeed, the release of butanal in the 3B168 sample, i.e., in the IL treated with copper. We can hypothesize that copper and copper particles may act as catalyzers for the further reaction of butene. Recently, several authors have reported the remarkable long-term stability and high selectivity towards alkenes of Cu nanoparticles as a promising alternative to replace precious-metal-based catalysts in selective hydrogenation [25]. It must be pointed out that even the concentration of butanal was too low to be detected by FTIR and NMR. These volatiles probably result from minor side reactions occurring during the degradation process.

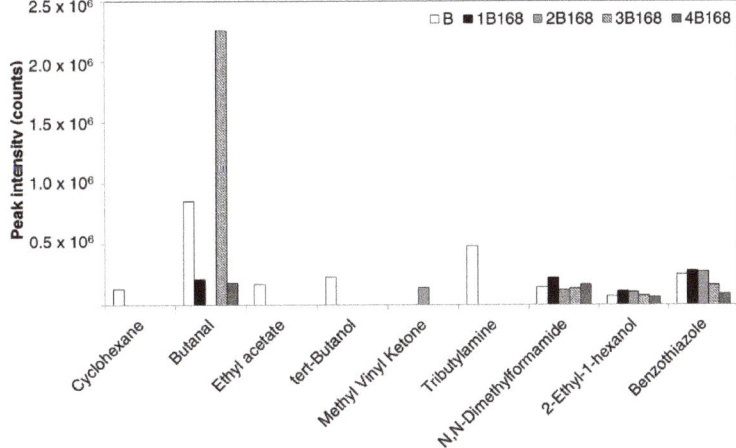

Figure 9. Relative intensity (counts) of 9 main compounds identified in samples B, 4B168, 1B168, 2B168 and 3B168 samples by HS-GCMS analysis.

4. Conclusions

The structural characterization of the ionic liquid [TBuMA][NTF$_2$], both fresh and after thermal treatment, with or without different metals, was carried out by HRMAS NMR and FTIR spectroscopic analyses, while HS-GC-MS was employed to reveal the formation of volatile compounds. The degradation products of [TBuMA][NTF$_2$] were characterized after thermal treatment at 200 °C for 4, 24 and 168 h in contact with AISI 304 steel, copper or brass, and without metals as a comparison. The combination of the above-mentioned techniques evidenced the occurrence of degradation processes of the cation. The data suggested a degradation mechanism compatible with the Hoffman elimination of one or two alkyl chains from the cation, with 1-butene being one of the degradation products after thermal treatment, both in the absence or presence of metal plates. The proposed multi-technique approach was revealed to be suitable for the characterization of the degradation compounds of [TBuMA][NTF$_2$] after thermal treatment in the presence of metals, thus proving to be a promising method for the selection of IL compounds that possess high stability and a suitable lifetime to meet the durability requirements of commercial and industrial solar thermal applications.

Supplementary Materials: The following supporting information can be downloaded at https://www.mdpi.com/article/10.3390/app12031652/s1: Figure S1: comparison of ^{1}H HRMAS spectrum of 4B168 (a,d) with 1D selective ^{1}H TOCSY obtained by irradiating H1,1 (9.16 ppm, b), H1,2 (8.31 ppm, c) and H1a,3 (5.13 ppm, e).; Figure S2: ^{1}H–^{13}C HSQC spectrum of 4B168; Table S1: assignment of ^{1}H and ^{13}C NMR signals of the ionic liquid B; Table S2: assignment of ^{1}H and ^{13}C NMR signals of the degradation compounds 1, 2 and 3.

Author Contributions: Conceptualization, A.F., S.P., E.B. (Emilia Bramanti) and A.L.; methodology, A.F., S.P., E.B. (Emilia Bramanti) and A.L.; investigation, F.N., S.P., E.B. (Emilia Bramanti), B.C., A.L. and E.B. (Enrico Berretti); writing—original draft preparation, A.F., S.P., E.B. (Emilia Bramanti); writing—review and editing, F.N., C.F., A.L., E.B. (Emilia Bramanti), S.P.; supervision, A.F., C.F.; funding acquisition, A.F., A.L. All authors have read and agreed to the published version of the manuscript.

Funding: This research was funded by the FELIX project (Fotonica ed Elettronica Integrate per l'Industria, project code no. 6455) and project MIUR Cluster CTN02_00018 «Energia» Codice progetto CTN02_00018_10016852 "NeMESi".

Conflicts of Interest: The authors declare no conflict of interest.

References

1. Welton, T. Ionic liquids: A brief history. *Biophys. Rev.* **2018**, *10*, 691–706. [CrossRef] [PubMed]
2. Paul, T.C.; Morshed, A.K.M.M.; Fox, E.B.; Visser, A.E.; Bridges, N.J.; Khan, J.A. Thermal performance of ionic liquids for solar thermal applications. *Exp. Therm. Fluid Sci.* **2014**, *59*, 88–95. [CrossRef]
3. Zhang, Z.; Salih, A.A.M.; Li, M.; Yang, B. Synthesis and characterization of functionalized ionic liquids for thermal storage. *Energy Fuels* **2014**, *28*, 2802–2810. [CrossRef]
4. Wadekar, V. V Ionic liquids as heat transfer fluids—An assessment using industrial exchanger geometries. *Appl. Therm. Eng.* **2017**, *111*, 1581–1587. [CrossRef]
5. Das, L.; Rubbi, F.; Habib, K.; Aslfattahi, N.; Saidur, R.; Baran Saha, B.; Algarni, S.; Irshad, K.; Alqahtani, T. State-of-the-art ionic liquid & ionanofluids incorporated with advanced nanomaterials for solar energy applications. *J. Mol. Liq.* **2021**, *336*, 116563.
6. Macfarlane, D.R.; Tachikawa, N.; Forsyth, M.; Pringle, J.M.; Howlett, P.C.; Elliott, G.D.; Davis, J.H.; Watanabe, M.; Simon, P.; Angell, C.A. Energy applications of ionic liquids. *Energy Environ. Sci.* **2014**, *7*, 232–250. [CrossRef]
7. Valkenburg, M.E.V.; Vaughn, R.L.; Williams, M.; Wilkes, J.S. Thermochemistry of ionic liquid heat-transfer fluids. *Thermochim. Acta* **2005**, *425*, 181–188. [CrossRef]
8. Salgado, J.; Parajó, J.J.; Fernández, J.; Villanueva, M. Long-term thermal stability of some 1-butyl-1-methylpyrrolidinium ionic liquids. *J. Chem. Thermodyn.* **2014**, *74*, 51–57. [CrossRef]
9. Monteiro, B.; Maria, L.; Cruz, A.; Carretas, J.M.; Marçalo, J.; Leal, J.P. Thermal stability and specific heats of coordinating ionic liquids. *Thermochim. Acta* **2020**, *684*, 178482. [CrossRef]
10. Liu, J.; Wang, F.; Zhang, L.; Fang, X.; Zhang, Z. Thermodynamic properties and thermal stability of ionic liquid-based nanofluids containing graphene as advanced heat transfer fluids for medium-to-high-temperature applications. *Renew. Energy* **2014**, *63*, 519–523. [CrossRef]
11. Huddleston, J.G.; Visser, A.E.; Reichert, W.M.; Willauer, H.D.; Broker, G.A.; Rogers, R.D. Characterization and comparison of hydrophilic and hydrophobic room temperature ionic liquids incorporating the imidazolium cation. *Green Chem.* **2001**, *3*, 156–164. [CrossRef]
12. Huang, Y.; Chen, Z.; Crosthwaite, J.M.; Aki, S.N.V.K.; Brennecke, J.F. Thermal stability of ionic liquids in nitrogen and air environments. *J. Chem. Thermodyn.* **2021**, *161*, 106560. [CrossRef]
13. Kosmulski, M.; Gustafsson, J.; Rosenholm, J.B. Thermal stability of low temperature ionic liquids revisited. *Thermochim. Acta* **2004**, *412*, 47–53. [CrossRef]
14. Maton, C.; De Vos, N.; Stevens, C.V. Ionic liquid thermal stabilities: Decomposition mechanisms and analysis tools. *Chem. Soc. Rev.* **2013**, *42*, 5963–5977. [CrossRef] [PubMed]
15. Perissi, I.; Bardi, U.; Caporali, S.; Fossati, A.; Lavacchi, A.; Vizza, F. Ionic liquids: Electrochemical investigation on corrosion activity of ethyl-dimethyl-propylammonium bis(trifluoromethylsulfonyl)imide at high temperature. *Russ. J. Electrochem.* **2012**, *48*, 434–441. [CrossRef]
16. Perissi, I.; Bardi, U.; Caporali, S.; Fossati, A.; Lavacchi, A. Ionic liquids as diathermic fluids for solar trough collectors' technology: A corrosion study. *Sol. Energy Mater. Sol. Cells* **2008**, *92*, 510–517. [CrossRef]
17. Verma, C.; Ebenso, E.E.; Quraishi, M.A.; Hussain, C.M. Recent developments in sustainable corrosion inhibitors: Design, performance and industrial scale applications. *Mater. Adv.* **2021**, *2*, 3806–3850. [CrossRef]
18. Perissi, I.; Bardi, U.; Caporali, S.; Lavacchi, A. High temperature corrosion properties of ionic liquids. *Corros. Sci.* **2006**, *48*, 2349–2362. [CrossRef]

19. Pisarova, L.; Gabler, C.; Dörr, N.; Pittenauer, E.; Allmaier, G. Thermo-oxidative stability and corrosion properties of ammonium based ionic liquids. *Tribol. Int.* **2012**, *46*, 73–83. [CrossRef]
20. Cao, Y.; Mu, T. Comprehensive Investigation on the Thermal Stability of 66 Ionic Liquids by Thermogravimetric Analysis. *Ind. Eng. Chem. Res.* **2014**, *53*, 8651–8664. [CrossRef]
21. Alam, M.T.; Jenkins, E.J. HR-MAS NMR Spectroscopy in Material Science. In *Advanced Aspects of Spectroscopy*; Farrukh, M.A., Ed.; IntechOpen: London, UK, 2012; pp. 279–306.
22. Ilharco, L.M.; Garcia, A.R.; Hargreaves, E.C.; Chesters, M.A. Comparative reflection-absorption infrared spectroscopy study of the thermal decomposition of 1-hexene on Ru(0001) and on Pt(111). *Surf. Sci.* **2000**, *459*, 115–123. [CrossRef]
23. Sowmiah, S.; Srinivasadesikan, V.; Tseng, M.C.; Chu, Y.H. On the chemical stabilities of ionic liquids. *Molecules* **2009**, *14*, 3780–3813. [CrossRef] [PubMed]
24. Janssen, C.H.C.; Macías-Ruvalcaba, N.A.; Aguilar-Martínez, M.; Kobrak, M.N. Copper extraction using protic ionic liquids: Evidence of the Hofmeister effect. *Sep. Purif. Technol.* **2016**, *168*, 275–283. [CrossRef]
25. Totarella, G.; Beerthuis, R.; Masoud, N.; Louis, C.; Delannoy, L.; De Jongh, P.E. Supported Cu Nanoparticles as Selective and Stable Catalysts for the Gas Phase Hydrogenation of 1,3-Butadiene in Alkene-Rich Feeds. *J. Phys. Chem. C* **2021**, *125*, 366–375. [CrossRef] [PubMed]

Article

Competitive Detection of Volatile Compounds from Food Degradation by a Zinc Oxide Sensor

Lucio Bonaccorsi [1], Andrea Donato [1], Antonio Fotia [2,*], Patrizia Frontera [1] and Andrea Gnisci [3]

- [1] Department of Civil, Energy, Environment and Material Engineering, Mediterranea University of Reggio Calabria, Via Graziella Loc. Feo di Vito, 89124 Reggio Calabria, Italy; lucio.bonaccorsi@unirc.it (L.B.); andrea.donato@unirc.it (A.D.); patrizia.frontera@unirc.it (P.F.)
- [2] Department of Information Engineering, Infrastructures and Sustainable Energy, Mediterranea University of Reggio Calabria, Via Graziella Loc. Feo di Vito, 89124 Reggio Calabria, Italy
- [3] Department of Heritage, Architecture, Urbanism (PAU), Mediterranea University of Reggio Calabria, Via dell'Università 25, 89124 Reggio Calabria, Italy; andrea.gnisci@unirc.it
- * Correspondence: antonio.fotia@unirc.it

Abstract: During the phenomenon of food degradation, several volatile organic compounds are generally released. In particular, due to lipid oxidation in stored and packed meat, hexanal is formed as a typical decomposition product. Therefore, its detection can provide an important indication of the quality and conservation of meat. Unfortunately, the simultaneous release of other compounds, such as 1-pentanol and 1-octen-3-ol, during the first phase of the degradation process can have an undesirable effect on the detection of hexanal. In this work, a metal oxide (MOX) sensor based on zinc oxide (ZnO) was prepared and tested for possible use in the monitoring of low concentrations of hexanal. The sensor was expected to detect the target volatile with minimum interference from all the others, when released all at the same time. For this purpose, the ZnO sensor was exposed to both pure and different mixtures of vapors of the main competing organic compounds. Comparing the results of the mixtures to the response relating to pure hexanal, it was highlighted that the presence of 1-pentanol and 1-octen-3-ol decreases the response of the sensor to hexanal in terms of the eR/R_0 ratio, especially at low concentrations (5–10 ppm), while at 50 ppm, the sensor response was comparable with the hexanal quantity, proving that its detection was less affected at higher concentrations.

Keywords: zinc oxide; gas sensor; hexanal; 1-pentanol; 1-octen-3-ol; MOX

1. Introduction

Food quality evaluation, shelf-life determination, and raw product control are important fields in which conductometric gas sensors have been applied in recent years. In these sensors, the gas detection mechanism is based on the conductance variation of a semiconducting metal oxide (MOX) layer caused by the chemical/physical adsorption of the gas molecules on the MOX surface [1]. MOX sensors have been intensively studied and used for their low production cost, reduced size, and high sensitivity to a broad range of chemicals. Compared to typical chemical analytical techniques used in food industry, the lower accuracy of MOX sensors is compensated by a faster response, higher portability for field usage, and less expansive instrumentation. A possible application of MOX sensors in the food industry for meat quality control is the detection of hexanal, a typical decomposition product of meat formed during lipid oxidation in stored and packed meat products [2–4]. Hexanal, indeed, is considered an important indicator of the freshness status [5,6] of raw meat products. In a preliminary work, we compared three different MOX sensors for hexanal detection and we found that zinc oxide (ZnO), used as the detecting layer for resistive sensors, showed the best compromise between the sensing temperature (T = 523 K) and response amplitude [7,8]. Indeed, the ZnO resistive gas sensor was tested with a broad range of hexanal concentrations in different working conditions, especially in

terms of the working temperatures and relative humidity. However, although the suitability of the ZnO sensor for hexanal detection was proved, the influence of other factors needs to be considered. In particular, in real food monitoring conditions, the decomposition process leads to the formation of numerous volatiles even during the early stage and the sensor must have the ability to detect the target volatile with minimal interference from others. In other words, the sensor must show sensitivity and selectivity towards hexanal. In general, as known, selectivity is one of the major issues in conductometric sensors made of semiconductor oxides because the sensing mechanism based on the exchange of electrons on the sensor surface during the "oxidation" of the volatile molecules is not a selective process and similar compounds can oxidize at the same time [9]. Several alternatives to improve the selectivity and sensitivity of MOX sensors are under continuous development, for example, the synthesis of particular architectures, such as nanowires, nanotubes, and nanosheets [10], exploiting the effect of nanostructured materials with respect to the bulk form [11], or the preparation of conductometric sensors based on p/n-type heterojunctions [12]. Similarly, the development of electronic noses requires the implementation of different sensor arrays to improve the ability to discriminate between similar volatile compounds [9,13–15]. However, electron noses, which ensure the widest applicability, are complex and voluminous equipment that require trained specialists for their use. A single sensor or at most a combination of two makes the detection system easy to use even for unskilled operators and easily transportable, even if it is not usable for generic analyses but suitably developed for a specific application. From this point of view, a complete characterization of the sensor's behavior in the environment of interest is a fundamental step for the realization of the measurement system. Therefore, in this work, we studied the effect of the coexistence of the main volatiles formed during meat degradation on the response of a ZnO sensor to hexanal, which, to the best of our knowledge, has not been reported yet. In the first stage of the degradation process, 1-pentanol and 1-octen-3-ol are the two main compounds released together with hexanal that can unexpectedly and unwantedly interfere during the detection of the MOX sensor [4,6,16]. The ZnO sensor was initially exposed to the three pure compounds and subsequently to binary and ternary mixtures of the three compounds, simulating a degradation process, to compare the responses and evaluate the effect of the interference. The results showed that the interaction of the three volatiles on the ZnO surface interferes in the hexanal detection, reducing the sensor's sensitivity in comparison to the response to the vapors of the pure component. A comparative analysis of the transient response, however, showed that the electrical resistance of the ZnO sensor in the presence of hexanal vapors had a very rapid decrease that was not observed with the other compounds. Therefore, in a detection device based on a single ZnO sensor, the intensity of the signal, which is related to the vapor concentration, should not be the only parameter that is relied on, but the combination with the decrease rate of the sensor's resistance, which is distinctive of hexanal, has to be considered so to improve the reliability of the response.

2. Experimental

2.1. Sensor Preparation and Characterization

Zinc oxide powder was prepared dissolving zinc nitrate (0.68 M) in aqueous solution and hydrolyzing it with a solution of potassium carbonate (1 M) [8]. The precipitate was filtered, washed with deionized water, dried at 383 K for 12 h, and then calcinated at 673 K for 2 h in air.

The powder was characterized by XRD analysis, D2 Phaser, equipped with a heating chamber, Anton Paar HTK 1200N (Bruker, Billerica, MA, USA), in the 2θ range 15–65° (Cu Kα_1 = 1.54056 Å) and by scanning electron microscopy (SEM) (Phenom ProX, Deben, Suffolk, UK). To evaluate the thermal stability of the synthesized zinc oxide, X-ray diffractometry at increasing temperatures was carried out to collect powder patterns at 298–373–473–523–573–673 K.

The ZnO powder was mixed with ethanol in an ultrasonic bath. The alcoholic paste was painted, with a controlled thickness, on the surface of an alumina support (3 mm × 6 mm) with interdigitated Pt electrodes on one side and a Pt heater on the other. The sensor was annealed at 673 K for 2 h in air before the sensing tests to stabilize the microstructure of the film.

2.2. Sensing Experiments

Liquid solutions of the three compounds, hexanal > 98%, 1-pentanol > 99%, 1-octen-3-ol > 98% (Sigma Aldrich, St. Louis, MO, USA), were used to generate pure vapors and mixtures for the experiments.

For the detection tests, the sensor was placed in a stainless-steel cell and exposed to 100 sccm of the vapor mixture, consisting of dry air (the carrier) and organic compounds. Pure vapors were obtained by bubbling dry air in pure liquid organics maintained at 258 (±0.1) K by a refrigerated circulating bath. Different concentrations (5–10–50 ppm) of vapors were obtained by regulating air fluxes by a battery of mass flow controllers (Brooks Instruments, Hatfield, PA, USA). Similarly, binary and ternary mixtures of vapors were obtained by flowing dry air in liquid mixtures of compounds according to the composition reported in Table 1.

Table 1. The composition of liquid and vapor mixtures used in the sensing tests.

Mixtures	Composition (w%)			
	Liquid	Vapor	Liquid	Vapor
Hexanal/1-pentanol	70/30	75/25	95/5	98/2
Hexanal/1-octen-3-ol	70/30	75/25	95/5	98/2
Hexanal /1-pentanol/1-octen-3-ol	-	-	90/5/5	96/2/2

The gas concentrations of the pure compounds were calculated from the vapor pressures at T = 258 K estimated by the Antoine equation with coefficients from literature data [17–19]. The calculated values were compared with the evaporated quantities by periodically weighting the remaining liquid using an analytical balance. In the case of binary and ternary mixtures, the vapor concentrations were calculated using Raoult's law in the hypothesis of ideal mixtures:

$$x_i P_i^0 = y_i P$$

where x_i = molar fraction of the component *I* in the liquid mixture, P_i^0 = vapor pressure of the component *I*, y_i = molar fraction of the component *I* in the gas mixture, and *P* = the total (atmospheric) pressure.

The sensor's resistance data were collected in the four-point mode by an Agilent 34970A multimeter while a dual-channel power supplier instrument, Agilent E3632A (Agilent Technologies, Santa Clara, CA, USA) allowed the sensor's temperature to be controlled, as explained elsewhere [7].

3. Results

3.1. ZnO Characterization

In Figure 1a, an SEM image of the synthesized ZnO powder is shown. The mean particle size of the powder ranged from 0.1–0.5 µm. The XRD pattern of ZnO acquired at 298 K is shown in Figure 1b and confirms the typical hexagonal structure of wurtzite (cell parameters: a = 3.249 Å, c = 5.207 Å) with crystallites size = 40.9 nm calculated by the Debye–Scherrer equation. MOX sensors, however, generally have a working temperature higher than 298 K. Thus, an investigation of the possible transformation/transitions of ZnO in temperatures ranging from 298–673 K was considered of interest, with T = 523 K being the working temperature of the sensor used in all the experiments and T = 673 K being the annealing temperature of the synthesized powder before the sensing tests. The results of the X-ray diffractometry and increasing temperature are shown in Figure 1b. After heating the oxide powder to 673 K, the hexagonal structure of ZnO did not show crystallographic variations as evidenced by the X-ray pattern and no increase in the peaks' intensity or variation in the peaks' shape were observed (Figure 1b), demonstrating that the particle size did not change at the annealing temperature. The only consequence of the powder heating was the shift in the diffractometric peaks towards lower angles when the temperature was increased, as shown in the inset of Figure 1b for the peak at 2-Tetha ≈ 31.7°. This is an indication of the thermal expansion of the unit cell volume with the temperature. The thermal stability of the sensing layer is important because in the case of the sintering of oxide particles, the sensor response decreases with time [20–22]. SEM analysis of the ZnO powder after the annealing treatment at T = 673 K confirmed the previous result because no evidence of morphologic variations was observed (Figure 1a).

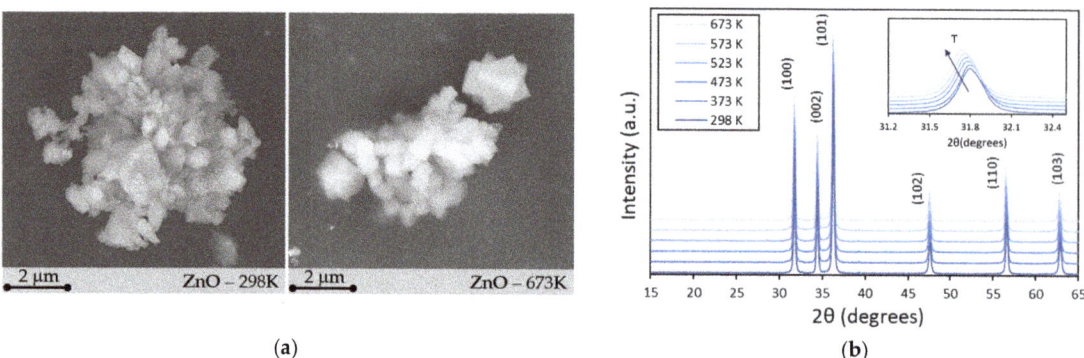

Figure 1. Electron microscopy (SEM) image of the ZnO powder as prepared at 298 K and after the annealing treatment at 673 K (**a**), X-ray patterns of ZnO from 298 K to 673 K and (inset) the temperature shift at 2θ = 31.7° (**b**).

3.2. Sensing Experiments

The sensor response was defined as the ratio of the reference resistance measured in dry air (R_0) and the resistance when the sensor was exposed to the vapor mixtures (R). In all sensing experiments the sensor temperature was maintained at 523 K because this working condition resulted in the highest response in a previous study on the ZnO sensitivity to hexanal [8]. The first set of experiments was carried out by measuring the response of pure compounds to different concentrations in dry air, as reported in Table 2 and shown in Figure 2. The lowest concentration was set at 5 ppm (in dry air) because this is considered the threshold value for the perception of hexanal by the human nose and therefore is used as an indication of the onset of meat degradation processes [23,24].

Table 2. ZnO sensor's response to pure compounds (T = 523 K).

Chemical Compound	Response (R_0/R)		
	C = 5 ppm	C = 10 ppm	C = 50 ppm
Hexanal	2.12	2.42	3.28
1-pentanol	1.30	1.68	2.60
1-octen-3-ol	1.50	2.10	4.10

Comparing the data in Table 2, it is evident that ZnO is responsive to all three organic vapors, although in a different way. The sensor's response to hexanal, as already reported in a previous work in which the ZnO response to different hexanal concentrations was studied [8], is the highest up to C = 10 ppm. However, the response to 1-octen-3-ol shows the faster rise and exceeds all others at C = 50 ppm (Figure 2a). In contrast, the ZnO sensitivity for 1-pentanol is the lowest at all concentrations. The high sensitivity of ZnO to hexanal at 5 ppm and the growth rate of the signal with the concentration (in Figure 2b) indicate that the sensor's surface at C = 50 ppm of hexanal is close to the saturation of the adsorption centers so that the increase in the response is close to reaching an asymptotic value [22–24]. The response of the MOX sensor is the result of a complex interaction between the oxide surface and the adsorbed gas molecules so that the electrical properties of the ZnO layer are modified due to the effect of the redox reactions taking place with the adsorbed molecules and the oxygen ions on the layer surface [20,21,25–30]. The organic substances considered in our tests are chemically different with different reagent groups, which makes the comparison of the sensor response more difficult: hexanal is an alkyl aldehyde, 1-octen-3-ol is an unsaturated secondary alcohol, and 1-pentanol is an aliphatic alcohol. In these cases, the mechanism of reaction involves the preferential interaction of hexanal with the ZnO surface, which occurs through the carbonyl group of the aldehyde [5,7], while in the case of alcohols, the main interaction occurs through the hydroxyl group. Both aldehyde and alcohol compounds interact with oxygen ions on the sensor's surface by oxidation reactions of the respective functional groups. Another important property that can affect the interaction at the interface between a vapor and the sensor's surface is the volatility of the interacting substance.

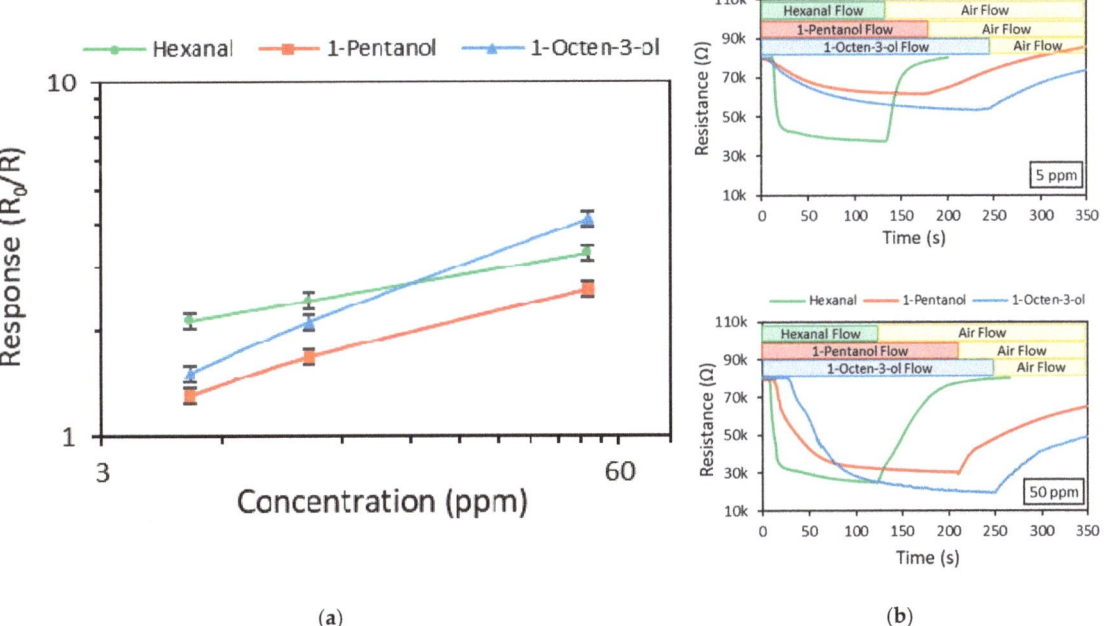

Figure 2. (**a**) Comparison of the sensor's response to pure compounds according to data from Table 2 and (**b**) the transient response of the ZnO sensor to pure volatiles at 5 (**top**) and 50 ppm (**bottom**).

Table 3 shows the boiling temperatures and vapor pressures of the three pure compounds for a comparison with the observed sensor responses. Hexanal, as expected, is the chemical compound that showed the lowest boiling temperature and highest vapor pressure; however, the sensor showed a higher response to 1-octen-3-ol than 1-pentanol despite the higher Tb and lower vapor pressure shown by unsaturated alcohol (Figure 2a).

For a more complete characterization of the ZnO sensor, the transient responses corresponding to the data shown in Table 2 are shown in Figure 2b for 2 concentrations: 5 and 50 ppm, respectively. The electrical resistance of the ZnO layer decreases with time when exposed to the organic vapors (n-type behavior) as a consequence of the redox reactions with the oxygen adsorbed on the sensor's surface and the consequent electron release into the conduction band of the semiconductor layer [7,20,31]. The transient response to hexanal compared to other volatiles was the fastest in the detection (R decrease), reaching the lowest resistance in seconds, and in regeneration in dry air (R increase) at all concentrations (Figure 2b). For the two alcohols, 1-pentanol and 1-octen-3-ol, the transient response was significantly slower than the hexanal response for both the detection and regeneration (Figure 2b). According to these results, although the value of the R_0/R ratio is the reference parameter used for evaluating the sensor's sensitivity, the response as a function of time can be important for discriminating among competing species, such as in the examples shown. During the degradation process, however, the alcohol vapors are released simultaneously with hexanal so the sensor's response cannot be considered a simple addition of the responses to pure compounds.

Table 3. Boiling temperatures and vapor pressures of pure compounds.

Chemical Compound	Tb [K]	P° (293 K) [kPa]	P° (258 K) [kPa]
Hexanal	402	2	1.1×10^{-1}
1-pentanol	411	5×10^{-1}	3.7×10^{-2}
1-octen-3-ol	447	3×10^{-2}	3.1×10^{-2}

To study the sensor's response in the presence of more than one volatile in the environment, a series of experiments were planned using binary and ternary mixtures. The results are shown in Table 4, where the composition of the vapor phases was determined as previously explained in the experimental paragraph.

Table 4. Sensor's response to binary and ternary mixtures.

| Vapor Mixture | Response (R_0/R) | | | |
	Composition (w%)	C = 5 ppm	C = 10 ppm	C = 50 ppm
Hexanal/1-pentanol	98/2	1.10	1.69	3.90
	75/25	1.20	1.53	2.74
Hexanal/1-octen-3-ol	98/2	1.55	2.25	5.50
	75/25	1.11	1.71	4.50
Hexanal /1-pentanol/1-octen-3-ol	96/2/2	1.10	1.50	2.85

According to [2,3,9,23,32], the beginning of food degradation is mainly correlated to the production of hexanal, which is predominant among the other organic compounds released and, indeed, is used as an indicator of the freshness of food and shelf-life determination. In our tests, this condition was simulated by binary mixtures with a concentration of 98/2 (w%). However, for a more complete characterization, the sensor was also tested with mixtures 75/25 (w%), with a higher concentration of the "competing" components (Table 4). Under the experimental conditions adopted, the vapors were generated by the liquid phase at T = 258 K and any reaction between the components was inhibited. The vapors were, then, conveyed directly to the sensor's surface, the temperature of which was maintained at 523 K. The sensing mechanism is mainly based on the oxidation reactions that are dominant on the sensor's surface, where, even in this case, due to the experimental conditions adopted, reactions between gas components are unlikely [33]. Indeed, the trend of the sensor's response and the stability of its working temperature confirmed the absence of side reactions due to the formation of intermediate compounds (such as acetal and hemiacetal) in the gas phase [33]. However, with respect to pure substances, different signals were registered when the binary and ternary mixtures were considered. Therefore, due to the complexity of the phenomena involved, which cannot only be described by a superposition of the signals from pure components, studying the sensor's response to binary and ternary mixtures was of fundamental importance. The sensor's response to hexanal/1-pentanol vapors compared to pure components is shown in Figure 3b. According to the data shown in Table 4, the sensor's behavior is strongly influenced by the presence of alcohol, as also evident in the comparison shown in Figure 3a. When exposed to the mixture 98/2, the sensor was more sensitive to 1-penthanol at low concentrations (5 and 10 ppm) and only when the compounds' concentration increased to 50 ppm did hexanal detection become predominant (Figure 3a). In the presence of a higher concentration of 1-pentanol, mixture 75/25, the dominant effect of alcohol was confirmed as the detection of the hexanal was almost completely inhibited and the trend of the signal was similar to the case of the pure compound (Figure 3a). For a more complete characterization of the mixture 98/2, Figure 3b shows the sensor's resistance as a function of time at C = 5 and 50 ppm. At low concentrations, the transient response follows the 1-pentanol trend while at 50 ppm, the resistance's variation is similar to that of pure hexanal, in agreement with the results shown in Figure 3a. The sensor's response to hexanal was only evident at the highest concentration.

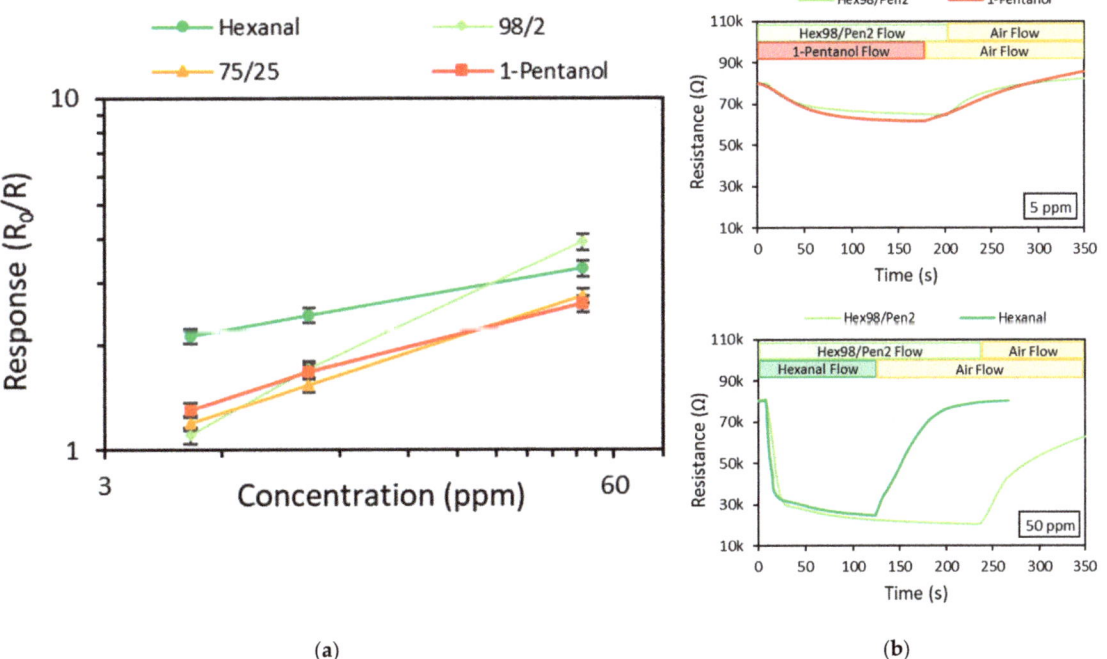

(a)   (b)

Figure 3. (a) Sensor's response to hexanal/1-pentanol mixtures (data from Table 4) and (b) transient response to mixture 98/2 (hexanal/1-pentanol) at 5 (**top**) and 50 ppm (**bottom**).

The effect of 1-octen-3-ol in combination with hexanal is shown in Figure 4 together with the response to the pure components. The ZnO sensor was significantly influenced by the presence of secondary unsaturated alcohol in both mixtures, 98/2 and 75/25, as evident from the data shown in Table 4 plotted in Figure 4a. The sensor signal, R_0/R, is affected by the alcohol even in the mixture 98/2 at all concentrations, from 5 to 50 ppm (Figure 4a). At the highest concentration, the response was even more intense than the sum of the two signals. The transient response shown in Figure 4b confirms that the sensor is mainly influenced by the 1-octen-3-ol traces at all concentrations, although after increasing the concentration of the mixture 98/2 to C = 50 ppm, the sensor's resistance showed a decreasing rate regarding the detection, which is similar to the response to pure hexanal, which was much faster than in the presence of pure 1-octen-3-ol.

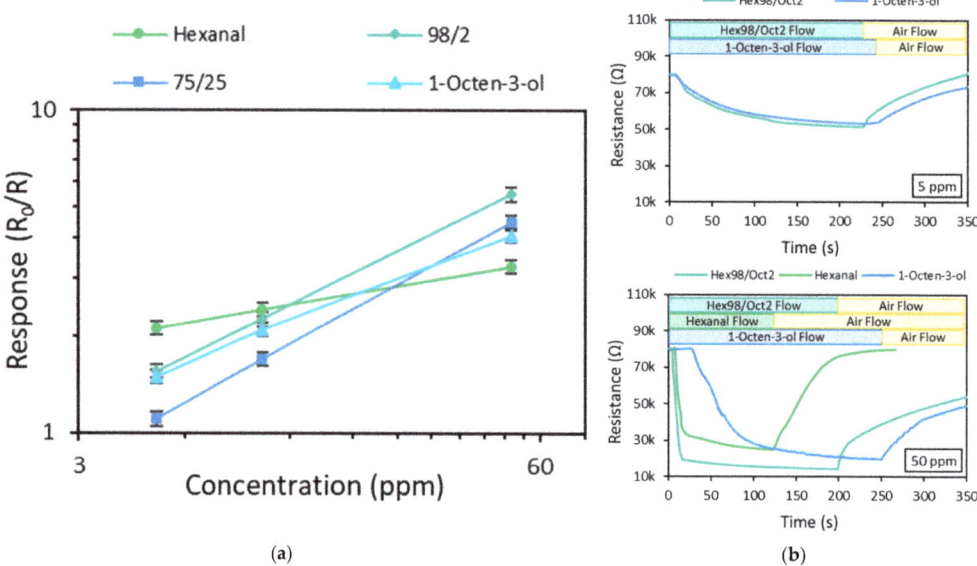

Figure 4. (**a**) Sensor's response to hexanal/1-octen-3-ol mixtures (data from Table 4) and (**b**) transient response to mixture 98/2 (hexanal/1-octen-3-ol mixtures) at 50 (**top**) and 50 ppm (**bottom**).

Finally, the ternary mixture with the composition 96/2/2 was used to study the ZnO sensor in an environment similar to the condition that develops when monitoring food degradation. Figure 5 shows the ZnO response to the different concentrations in dry air compared to the response recorded in the detection of hexanal vapors only.

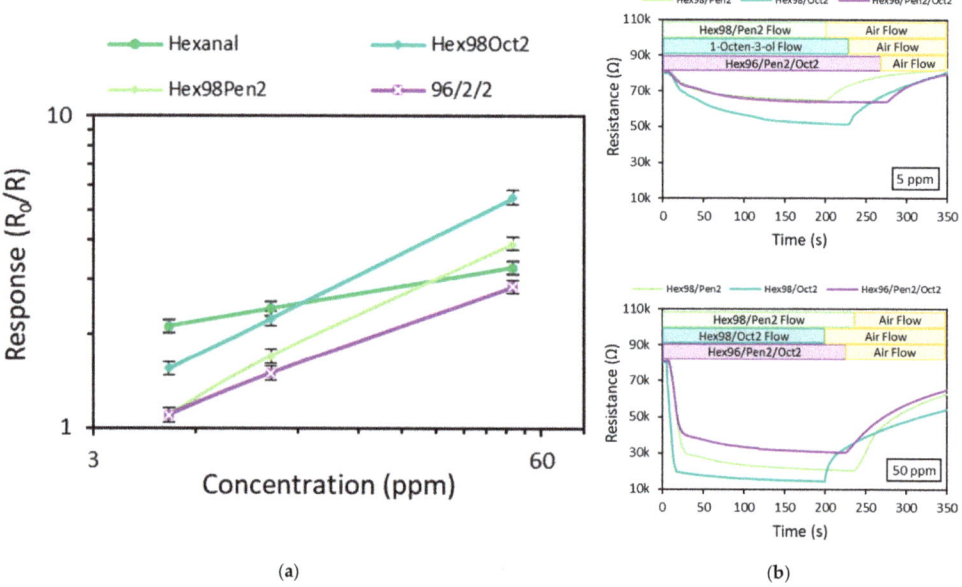

Figure 5. (**a**) Sensor's response to the ternary mixture 96/2/2 (data from Table 4) and (**b**) the transient response of the ZnO sensor to the ternary mixture 96/2/2 for 5 (**top**) and 50 ppm (**bottom**).

Although the response increased with the concentration, the sensor was clearly negatively affected by the other two components of the mixture and low sensitivity was observed, especially at low concentrations. The comparison with the responses to the binary mixtures (Figure 5a) shows that the inhibition effect on the hexanal detection is mainly due to the presence of 1-pentanol vapors, which is the dominant component among the two alcohols.

A similar conclusion can be obtained from the analysis of the transient responses for the three mixtures tested, as shown in Figure 5b. At 5 ppm, the sensor's resistance decreases over time, showing a trend that is very similar to the transient response of pure 1-pentanol (Figure 5b top). After increasing the concentration to 50 ppm, it was observed that the sensor's resistance trend over time (Figure 5b bottom) is not sufficiently different from the transient responses to binary mixtures to be usable for the identification of the compounds in the mixture, although the presence of hexanal vapors is recognizable by the steep resistance decrease.

4. Conclusions

In the work presented, a resistive sensor based on ZnO oxide was prepared and tested for possible use in the monitoring of food degradation phenomena. The ZnO sensor showed a good sensitivity to low concentrations of hexanal, the target compound used in this work as an indicator of meat degradation. In a realistic scenario, the sensor must be able to discriminate among other organic volatiles released at the same time, so tests of the sensitivity were carried out by exposing the sensor to different mixture of vapors of the main competing organic compounds, 1-pentanol and 1-octen-3-ol. In this condition, the sensitivity to hexanal was reduced in the range of low concentrations (5–10 ppm) while at 50 ppm, the sensor's response was comparable with the hexanal quantity. The experiments with binary and ternary mixtures showed, however, that the transient response of the ZnO sensor was significantly influenced by the presence of hexanal in the vapor mixture at all the concentrations tested. Although not so evident in terms of the intensity of the signal R_0/R, the detection of hexanal in the vapor mixtures was characterized by a steep decrease with time of the sensor's resistance even at low concentrations, which was not shown by the other organic components.

The ZnO sensor belongs to the family of conductometric sensors and for this reason, it is affected by some common limitations, such as reduced selectivity, electrical drift, and response reproducibility. The results obtained in this study indicate the possible use of ZnO for the monitoring of food degradation; however, further characterization of the material behavior is needed, and durability and repeatability studies, to confirm the observed potential regarding its use as a sensor.

Author Contributions: Conceptualization, L.B., A.D. and P.F.; methodology, A.F., A.G. and L.B.; formal analysis, L.B. and A.F.; investigation, A.F. and A.G.; data curation, A.F. and A.G.; writing—original draft preparation, A.F. and A.G.; writing—review and editing, L.B. and P.F.; supervision L.B., A.D. and P.F. All authors have read and agreed to the published version of the manuscript.

Funding: This research was funded by Regione Calabria, POR Calabria FESR/FSE 2014-2020: Asse 12 Azione 10.5.6 "Mobilità internazionale di dottorandi".

Conflicts of Interest: The authors declare no conflict of interest.

References

1. Gnisci, A.; Fotia, A.; Bonaccorsi, L.; Donato, A. Effect of Working Atmospheres on the Detection of Diacetyl by Resistive SnO$_2$ Sensor. *Appl. Sci.* **2022**, *12*, 367. [CrossRef]
2. Shahidi, F.; Pegg, R.B. Hexanal As an Indicator of Meat Flavor Deterioration. *J. Food Lipids* **1994**, *1*, 177–186. [CrossRef]
3. Azarbad, M.H.; Jeleń, H. Determination of Hexanal—an Indicator of Lipid Oxidation by Static Headspace Gas Chromatography (SHS-GC) in Fat-Rich Food Matrices. *Food Anal. Methods* **2015**, *8*, 1727–1733. [CrossRef]
4. Schindler, S.; Krings, U.; Berger, R.G.; Orlien, V. Aroma Development in High Pressure Treated Beef and Chicken Meat Compared to Raw and Heat Treated. *Meat Sci.* **2010**, *86*, 317–323. [CrossRef]

5. Abd Wahab, N.Z.; Nainggolan, I.; Nasution, T.I.; Derman, M.N.; Shantini, D. Highly Response and Sensitivity Chitosan-Polyvinyl Alcohol Based Hexanal Sensors. *MATEC Web. Conf.* **2016**, *78*, 01072. [CrossRef]
6. Jayasena, D.D.; Ahn, D.U.; Nam, K.C.; Jo, C. Flavour Chemistry of Chicken Meat: A Review. *Asian-Australas. J. Anim. Sci.* **2013**, *26*, 732–742. [CrossRef]
7. Malara, A.; Bonaccorsi, L.; Donato, A.; Frontera, P.; Piscopo, A.; Poiana, M.; Leonardi, S.G.; Neri, G. Sensing Properties of Indium, Tin and Zinc Oxides for Hexanal Detection. In *Sensors*; Andò, B., Baldini, F., di Natale, C., Ferrari, V., Marletta, V., Marrazza, G., Militello, V., Miolo, G., Rossi, M., Scalise, L., et al., Eds.; Lecture Notes in Electrical Engineering; Springer International Publishing: Cham, Switzerland, 2019; Volume 539, pp. 39–44. ISBN 978-3-030-04323-0.
8. Malara, A.; Bonaccorsi, L.; Donato, A.; Frontera, P.; Neri, G. *Doped Zinc Oxide Sensors for Hexanal Detection*; Springer International Publishing: Cham, Switzerland, 2020; Volume 629, ISBN 9783030375577.
9. Ponzoni, A.; Comini, E.; Concina, I.; Ferroni, M.; Falasconi, M.; Gobbi, E.; Sberveglieri, V.; Sberveglieri, G. Nanostructured Metal Oxide Gas Sensors, a Survey of Applications Carried out at SENSOR Lab, Brescia (Italy) in the Security and Food Quality Fields. *Sensors* **2012**, *12*, 17023–17045. [CrossRef]
10. Li, F.; Gao, X.; Wang, R.; Zhang, T.; Lu, G.; Barsan, N. Design of Core–Shell Heterostructure Nanofibers with Different Work Function and Their Sensing Properties to Trimethylamine. *ACS Appl. Mater. Interfaces* **2016**, *8*, 19799–19806. [CrossRef]
11. Faggio, G.; Gnisci, A.; Messina, G.D.S.; Lisi, N.; Capasso, A.; Lee, G.-H.; Armano, A.; Sciortino, A.; Messina, F.; Cannas, M.; et al. Carbon Dots Dispersed on Graphene/SiO_2/Si: A Morphological Study. *Phys. Status Solidi* **2019**, *216*, 1800559. [CrossRef]
12. Xu, Y.; Zheng, L.; Yang, C.; Liu, X.; Zhang, J. Highly Sensitive and Selective Electronic Sensor Based on Co Catalyzed SnO_2 Nanospheres for Acetone Detection. *Sens. Actuators B Chem.* **2020**, *304*, 127237. [CrossRef]
13. Wasilewski, T.; Gębicki, J.; Kamysz, W. Bioelectronic Nose: Current Status and Perspectives. *Biosens. Bioelectron.* **2017**, *87*, 480–494. [CrossRef] [PubMed]
14. Arshak, K.; Moore, E.; Lyons, G.M.; Harris, J.; Clifford, S. A Review of Gas Sensors Employed in Electronic Nose Applications. *Sens. Rev.* **2004**, *24*, 181–198. [CrossRef]
15. Song, L.; Yang, L.; Wang, Z.; Liu, D.; Luo, L.; Zhu, X.; Xi, Y.; Yang, Z.; Han, N.; Wang, F.; et al. One-Step Electrospun SnO_2/MO_x Heterostructured Nanomaterials for Highly Selective Gas Sensor Array Integration. *Sens. Actuators B Chem.* **2019**, *283*, 793–801. [CrossRef]
16. Soncin, S.; Chiesa, L.M.; Cantoni, C.; Biondi, P.A. Preliminary Study of the Volatile Fraction in the Raw Meat of Pork, Duck and Goose. *J. Food Compos. Anal.* **2007**, *20*, 436–439. [CrossRef]
17. Covarrubias-Cervantes, M.; Mokbel, I.; Champion, D.; Jose, J.; Voilley, A. Saturated Vapour Pressure of Aroma Compounds at Various Temperatures. *Food Chem.* **2004**, *85*, 221–229. [CrossRef]
18. Palczewska-Tulinska, M.; Oracz, P. Vapor Pressures of Hexanal, 2-Methylcyclohexanone, and 2-Cyclohexen-1-One. *J. Chem. Eng. Data* **2006**, *51*, 639–641. [CrossRef]
19. Available online: Https://www.Cheric.Org/ (accessed on 18 November 2021).
20. Dey, A. Semiconductor Metal Oxide Gas Sensors: A Review. *Mater. Sci. Eng. B Solid-State Mater. Adv. Technol.* **2018**, *229*, 206–217. [CrossRef]
21. Oprea, A.; Degler, D.; Barsan, N.; Hemeryck, A.; Rebholz, J. *Basics of Semiconducting Metal Oxide–Based Gas Sensors*; Elsevier Inc.: Amsterdam, The Netherlands, 2018; ISBN 9780128112243.
22. Wang, C.; Yin, L.; Zhang, L.; Xiang, D.; Gao, R. Metal Oxide Gas Sensors: Sensitivity and Influencing Factors. *Sensors* **2010**, *10*, 2088–2106. [CrossRef]
23. Brewer, M.S.; Vega, J.D. Detectable Odor Thresholds of Selected Lipid Oxidation Compounds in a Meat Model System. *J. Food Sci.* **1995**, *60*, 592–595. [CrossRef]
24. Ajuyah, A.O.; Fenton, T.W.; Hardin, R.T.; Sim, J.S. Measuring Lipid Oxidation Volatiles in Meats. *J. Food Sci.* **1993**, *58*, 270–273. [CrossRef]
25. Schmidt, O.; Kiesel, P.; van de Walle, C.G.; Johnson, N.M.; Nause, J.; Dohler, G.H. Effects of an Electrically Conducting Layer at the Zinc Oxide Surface. *Jpn. J. Appl. Phys.* **2005**, *44*, 7271–7274. [CrossRef]
26. Schmidt, O.; Geis, A.; Kiesel, P.; van de Walle, C.G.; Johnson, N.M.; Bakin, A.; Waag, A.; Döhler, G.H. Analysis of a Conducting Channel at the Native Zinc Oxide Surface. *Superlattices Microstruct.* **2006**, *39*, 8–16. [CrossRef]
27. Janotti, A.; van de Walle, C.G. Fundamentals of Zinc Oxide as a Semiconductor. *Rep. Prog. Phys.* **2009**, *72*, 126501. [CrossRef]
28. Korotcenkov, G. Metal Oxides for Solid-State Gas Sensors: What Determines Our Choice? *Mater. Sci. Eng. B Solid-State Mater. Adv. Technol.* **2007**, *139*, 1–23. [CrossRef]
29. Barsan, N.; Weimar, U. Understanding the Fundamental Principles of Metal Oxide Based Gas Sensors; the Example of CO Sensing with SnO_2 Sensors in the Presence of Humidity. *J. Phys. Condens. Matter* **2003**, *15*, R813–R839. [CrossRef]
30. Barsan, N.; Weimar, U. Conduction Model of Metal Oxide Gas Sensors. *J. Electroceram.* **2001**, *7*, 143–167. [CrossRef]
31. Hjiri, M.; el Mir, L.; Leonardi, S.G.; Pistone, A.; Mavilia, L.; Neri, G. Al-Doped ZnO for Highly Sensitive CO Gas Sensors. *Sens. Actuators B Chem.* **2014**, *196*, 413–420. [CrossRef]
32. Galstyan, V.; Bhandari, M.; Sberveglieri, V.; Sberveglieri, G.; Comini, E. Metal Oxide Nanostructures in Food Applications: Quality Control and Packaging. *Chemosensors* **2018**, *6*, 16. [CrossRef]
33. Agirre, I.; Barrio, V.L.; Güemez, M.B.; Cambra, J.F.; Arias, P.L. Acetals as possible diesel additives. *Econ. Eff. Biofuel Prod.* **2011**, *15*, 299–316.

Article

Light Harvesting in Silicon Nanowires Solar Cells by Using Graphene Layer and Plasmonic Nanoparticles

Ali Elrashidi [1,2]

[1] Department of Electrical Engineering, University of Business and Technology, Jeddah 21432, Saudi Arabia; a.elrashidi@ubt.edu.sa
[2] Department of Engineering Physics, Alexandria University, Alexandria 21544, Egypt

Abstract: In this work, a silicon nanowire solar cell for efficient light harvesting in the visible and near-infrared regions is introduced. In this structure, the silicon nanowires (SiNWs) are coated with a graphene layer and plasmonic nanoparticles are distributed on the top surface of the silicon substrate layer. The proposed structure is simulated using the finite difference time domain (FDTD) method to determine the performance of the solar cell by calculating the open-circuit voltage, fill factor, short-circuit current density, and power conversion efficiency. The absorbed light energy is compared for different nanoparticle materials, namely Au, Ag, Al, and Cu, and Au NPs give the best performance. Different values of the radius of the Au NP are simulated, namely 30, 40, 50, and 60 nm, to determine the optimum radius, and the effect of excess carrier concentration on the solar cell performance is also tested. The obtained open-circuit voltage is 0.63 V, fill factor is 0.73, short-circuit current density is 41.7 mA/cm^2, and power conversion efficiency is 19.0%. The proposed SiNW solar cell improves the overall efficiency by almost 60%. Furthermore, the effects of the NW length and distance between NWs are also studied in this work. Finally, the distribution of the optical power in different layers along the solar cell and for different solar cell structures is also illustrated in this paper.

Keywords: plasmonic nanoparticles; silicon solar cell; graphene; short-circuit current density; open-circuit voltage; power conversion efficiency

1. Introduction

Crystalline silicon nanowires (SiNWs) usually exhibit a highly efficient light absorption and an antireflection of the incident light greater than that of a flat surface silicon material of the same size [1]. Geometrical scattering and antenna effect of the SiNWs are the main reasons for high light-harvesting performance [2,3], where the NW density and diameter determine the absorbed wavelength, which is called the resonance wavelength [4]. One more advantage of the SiNW solar cell is the simplicity of the fabrication process with a large-area scalable structure, as introduced by Garnett and Yang [5]. This process produced a high open-circuit voltage value, V_{oc}, and high fill factor, FF, which lead to producing higher overall efficiency, μ, of the solar cell for high values of short-circuit current density, J_{sc}. However, we still need to increase the value of the J_{sc}, which can be obtained by using an absorber layer, such as graphene, C [6]. For its good electric properties, as a thin and highly transparent material, graphene has been greatly used in recent research [7]. Graphene on a silicon surface forms a Schottky junction solar cell with a very low efficiency, which can be improved by engineering the interface with a passivation layer to increase the graphene work function [8]. On the other hand, plasmonic nanoparticles have high absorption properties which depend on the surrounding medium and the size, shape, and material of the nanoparticles [9]. In this work, we used a graphene layer and plasmonic NPs distributed on the top of a silicon surface to enhance the power conversion efficiency (PCE) of the SiNW solar cell [10–12].

Large-area SiNWs with radial PN junctions have been produced using a simple fabrication method by Garnett and Yang [5]. The fabricated solar cell was based on a room-temperature aqueous etching method and low-temperature thin film deposition with a rapid thermal annealing crystallization process. The given method increased the PCE to be 5.3% with V_{oc} = 0.56 V, FF = 0.607, and J_{sc} = 17.32 mA/cm^2.

A SiNW solar cell grown on a silicon base and covered with an on-site graphene layer for Schottky junction was introduced by Wallace et al. [6]. The graphene layer increased the carrier collection and enhanced the solar cell efficiency to be 3.83%. The obtained open-circuit voltage is 0.47 V, the fill factor is 0.59, and the short-circuit current density is 13.7 mA/cm^2. Linwei Yu and Pere Cabarrocas fabricated a SiNW nanostructure on a silicon base photovoltaic, which achieved light absorption [1]. In addition, a 3D radial junction solar cell has been developed in order to overcome light-induced degradation. The experimental results illustrated that the maximum obtained open-circuit voltage from the solar cell is 0.82 V and the FF is 0.73, while the short-circuit current density is 15.2 mA/cm^2, which leads to a power conversion efficiency of 9.3%. A large-scale silicon solar microcell fabricated from bulk wafers using transferring printing technique was presented by Yoon et al. [13]. The produced device has many features such as mechanical flexibility, transparency, and ultrathin microconcentrator designs. Theoretical and experimental electrical, mechanical, and optical characteristics of several types of modules were illustrated in this work. The measured V_{oc} is 0.51, FF is 0.61, J_{sc} is 33.6 mA/cm^2, and PCE is 11.6%.

Augusto et al. showed the analytical model of the recombination mechanisms of a silicon solar cell with low bandgap-voltage offset [14]. The work showed the effect of changing excess carrier density on the effective lifetime of the carriers. The untextured structure, with 50 μm thickness, produced 0.764 V as an open-circuit voltage, a fill factor of 0.86, and a short-circuit current density of 38.8 mA/cm^2.

Moreover, the surface plasmon localization (SPL) technique on plasmonic nanoparticles is used as an efficient method to improve the optical absorption of solar cells, metal–dielectric–metal, and metamaterial applications [15–19]. The NP shapes and materials are the main parameters that affect the solar cell and sensor performance [20–23].

In this paper, the light harvesting of SiNW solar cells has been improved by using a graphene layer and gold nanoparticles. Short-circuit current density, power conversion efficiency, and light absorption are simulated by applying a finite difference time domain method using the Lumerical FDTD solutions software package. The effect of changing excess carrier concentration on the PCE is also illustrated in this work. For different nanoparticle materials, the light absorption and PCE are illustrated as a function of the wavelengths. Moreover, the effects of NW length and separation between two successive NWs are also given in this work. In addition, the optical power distribution in different monitoring layers for different proposed solar cell structures is given in this work.

2. Analytical Model

The overall solar cell efficiency can be calculated using a single diode model and by applying the Green empirical expression [4]. Short-circuit current density, J_{sc}, can be calculated using Equation (1) by assuming that an incident photon will produce an electron [9].

$$J_{sc} = \frac{q}{hc} \int I(\lambda) A(\lambda) \lambda \, d\lambda \qquad (1)$$

where q is the electron charge, h is the Planck constant, c is the speed of light, $I(\lambda)$ is the standard air mass 1.5 (AM1.5) spectral irradiance, and $A(\lambda)$ is the optical absorption. On the other hand, the open-circuit voltage, V_{oc}, of silicon nanowire pillars is given by Equation (2) [24].

$$V_{OC} = V_{Th} \ln[\frac{\Delta n (N_A + \Delta n)}{n_i^2}] \qquad (2)$$

where V_{Th} is the thermal voltage which equals 25.8 mV, N_A is the doping concentration ($N_A = 3.3 \times 10^{17}$ cm^{-3}), Δn is the excess carrier concentration for n-type ($\Delta n = 10^{15}$ cm^{-3}) [11], and n_i is the intrinsic carrier concentration. The open-circuit voltage of the silicon nanowires can be calculated and considered as a fixed value equal to 0.626 V as it depends on the energy level of the material and for $n_i = 10^{10}$ cm^{-3} at room temperature.

Hence, the fill factor can be calculated using Equation (3) [9].

$$FF = \frac{\frac{V_{oc}}{V_{Th}} - \ln\left(\frac{V_{oc}}{V_{Th}} + 0.72\right)}{\frac{V_{oc}}{V_{Th}} + 1} \quad (3)$$

The maximum output power, P_{max}, can be calculated using Equation (4).

$$P_{max} = J_{sc} \times V_{oc} \times FF \quad (4)$$

Hence, the overall solar cell efficiency, η, can be calculated as a ratio of maximum output power to solar input power.

Gold, Au, plasmonic nanoparticles distributed on the substrate layer will change the optical power absorbed in the active layer and depend on the maximum value of reflectivity. Nanoparticle shape has a great effect on the transmitted optical power as well as the relative permittivity of the gold nanoparticles and the dielectric function of the surrounding medium [9]. The maximum reflectivity can be calculated using Equation (5).

$$\lambda_{max} = \frac{L}{g}\left(\frac{\varepsilon_{Au}\varepsilon_m(\lambda_{max})}{\varepsilon_m + \varepsilon_{Au}(\lambda_{max})}\right)^{1/2} \quad (5)$$

where ε_m is the permittivity of the surrounding medium, ε_{Au} is a gold nanoparticle dielectric constant at corresponding λ_{max}, g is an integer, and L is structural periodicity. Hence, the dielectric permittivity can be expressed by using a multioscillator Drude–Lorentz model [9] as given in Equation (6):

$$\varepsilon_{Au} = \varepsilon_\infty - \frac{\omega_D^2}{\omega^2 + j\omega\gamma_D} - \sum_{k=1}^{6} \frac{\delta_k \omega_k^2}{\omega^2 - \omega_k^2 + 2j\omega\gamma_k} \quad (6)$$

where ε_∞ is the gold high-frequency dielectric permittivity, ω_D and γ_D are the plasma and collision frequencies of the free electron gas, δ_k is the amplitude of Lorentz oscillator, ω_k is the resonance angular frequencies, and γ_k is the damping constants for k value from 1 to 6.

3. Proposed Structure

In this work, we used an electromagnetic wave solver, Lumerical FDTD solutions software, to design and analyze the proposed structure. A unit cell of the proposed structure has been simulated using the FDTD method. The overall dimensions of the unit cell are 500 × 500 × 700 nm^3 in the three dimensions x, y, and z, respectively. A silicon substrate with height h_2 = 3 µm is placed on the top of a back oxide material with height h_1 = 1 µm, as illustrated in Figure 1. Silicon nanowires are grown on the top of the silicon substrate with height H = 5 µm and period P = 550 nm between each NW. In addition to that, a graphene layer can be used on the top of substrate material and on the NWs as well.

The boundary conditions are considered as periodic structures in x-direction and y-direction, as the structure is extended in both x- and y-dimensions, and perfect matching layer in z-direction, as the structure is not repeated in z-direction, where the minimum mesh size is 20 nm in all directions. A plane wave source with wavelength band 400–1100 nm and offset time 7.5 fs is used as a light source. In addition, the solar generation calculation region is given in the active layer to calculate the short-circuit current density.

The refractive index of back oxide is 1.4; however, the value for silicon is a function of the wavelength and follows the Aspnes and Studna model [25], and the refractive index of graphene follows the Phillip and Taft model [26]. On the other hand, the refractive index of plasmonic NPs is summarized using Equation (6) in Table 1 [27].

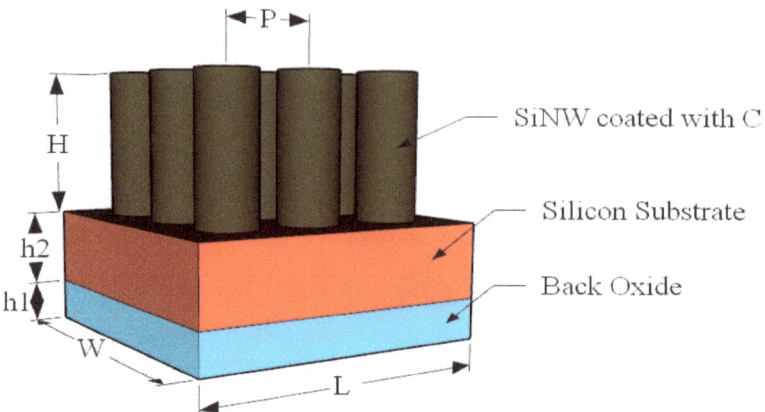

Figure 1. Schematic diagram of the proposed structure.

Table 1. Plasmonic parameters used for the metallic nanoparticle.

Material	Term	Strength	Plasma Frequency	Resonant Frequency	Damping Frequency
Ag	0	0.8450	$0.136884 \times 10^{+17}$	$0.000000 \times 10^{+00}$	$0.729239 \times 10^{+14}$
	1	0.0650	$0.136884 \times 10^{+17}$	$0.123971 \times 10^{+16}$	$0.590380 \times 10^{+16}$
	2	0.1240	$0.136884 \times 10^{+17}$	$0.680775 \times 10^{+16}$	$0.686701 \times 10^{+15}$
	3	0.0110	$0.136884 \times 10^{+17}$	$0.124351 \times 10^{+17}$	$0.987512 \times 10^{+14}$
	4	0.8400	$0.136884 \times 10^{+17}$	$0.137993 \times 10^{+17}$	$0.139163 \times 10^{+16}$
	5	5.6460	$0.136884 \times 10^{+17}$	$0.308256 \times 10^{+17}$	$0.367506 \times 10^{+16}$
Au	0	0.7600	$0.137188 \times 10^{+17}$	$0.000000 \times 10^{+00}$	$0.805202 \times 10^{+14}$
	1	0.0240	$0.137188 \times 10^{+17}$	$0.630488 \times 10^{+15}$	$0.366139 \times 10^{+15}$
	2	0.0100	$0.137188 \times 10^{+17}$	$0.126098 \times 10^{+16}$	$0.524141 \times 10^{+15}$
	3	0.0710	$0.137188 \times 10^{+17}$	$0.451065 \times 10^{+16}$	$0.132175 \times 10^{+16}$
	4	0.6010	$0.137188 \times 10^{+17}$	$0.653885 \times 10^{+16}$	$0.378901 \times 10^{+16}$
	5	4.3840	$0.137188 \times 10^{+17}$	$0.202364 \times 10^{+17}$	$0.336362 \times 10^{+16}$
Cu	0	0.5750	$0.164535 \times 10^{+17}$	$0.000000 \times 10^{+00}$	$0.455775 \times 10^{+14}$
	1	0.0610	$0.164535 \times 10^{+17}$	$0.442101 \times 10^{+15}$	$0.574276 \times 10^{+15}$
	2	0.1040	$0.164535 \times 10^{+17}$	$0.449242 \times 10^{+16}$	$0.160433 \times 10^{+16}$
	3	0.7230	$0.164535 \times 10^{+17}$	$0.805202 \times 10^{+16}$	$0.488135 \times 10^{+16}$
	4	0.6380	$0.164535 \times 10^{+17}$	$0.169852 \times 10^{+17}$	$0.654037 \times 10^{+16}$
Al	0	0.5230	$0.227583 \times 10^{+17}$	$0.000000 \times 10^{+00}$	$0.714047 \times 10^{+14}$
	1	0.2270	$0.227583 \times 10^{+17}$	$0.246118 \times 10^{+15}$	$0.505910 \times 10^{+15}$
	2	0.0500	$0.227583 \times 10^{+17}$	$0.234572 \times 10^{+16}$	$0.474006 \times 10^{+15}$
	3	0.1660	$0.227583 \times 10^{+17}$	$0.274680 \times 10^{+16}$	$0.205251 \times 10^{+16}$
	4	0.0300	$0.227583 \times 10^{+17}$	$0.527635 \times 10^{+16}$	$0.513810 \times 10^{+16}$

4. Results and Discussion

To maximize the absorbed light in the visible and NIR regions, which leads to an increase in the J_{sc} and the overall frequency, different proposals will be produced. The absorption of a traditional SiNW solar cell will be compared to SiNWs coated with a 2D graphene layer and with/without Au plasmonic nanoparticles distributed on the silicon substrate layer. Figure 2a,b illustrates the absorption and transmission of the four proposed techniques. As shown in Figure 2a, the absorptions of the traditional SiNWs and of SiNWs coated with graphene are almost the same; however, a high absorption has been obtained when Au NPs are distributed on the top silicon substrate surface. The light transmitted through the solar cell is illustrated given in Figure 2b. However, it is very hard to distinguish between four cases, so the short-circuit current density and consequently PCE is the used in such case to choose which structure has better performance.

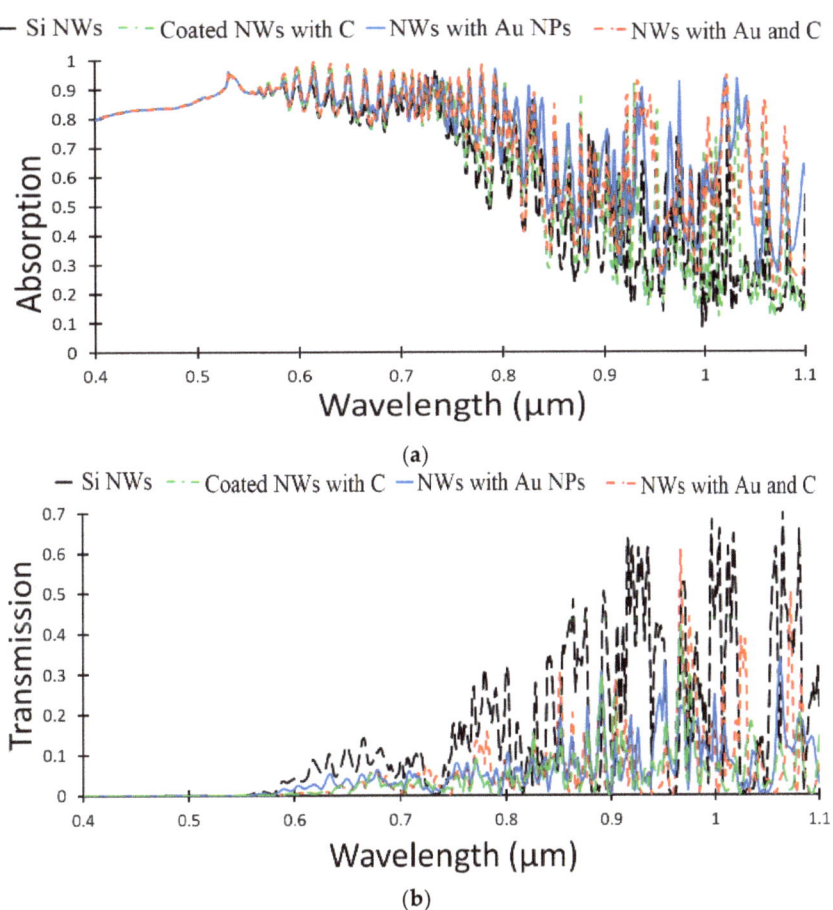

Figure 2. (a) Absorption and (b) transmission of the four proposed structures: traditional SiNWs, SiNWs coated with a graphene layer, SiNWs with Au NPs distributed on the substrate, and SiNWs with graphene layer and Au NPs.

The resultant short-circuit current density and the power conversion efficiency for each structure are given in Table 2. The maximum J_{sc}, 34.72 mA/cm^2, is obtained when the SiNWs coated with a graphene layer and Au NPs are distributed on the silicon substrate surface, which gives maximum overall efficiency, 15.8%, as shown in Table 2.

Table 2. Short-circuit current density and overall efficiency of the proposed four structures.

Proposed Structure	J_{sc} (mA/cm^2)	PCE (%)
Traditional SiNWs	31.56	14.40
SiNWs coated with graphene layer	31.91	14.60
SiNWs with Au NPs	33.20	15.10
SiNWs with graphene layer and Au NPs	34.72	15.80

The absorbed light is increased and the short-circuit current density and power conversion efficiency are consequently increased as the graphene layer is added, due to its high absorption properties. On the other hand, adding Au NPs increased the absorbed light according to Equations (5) and (6). To compare between the given two structures, SiNWs with Au NPs and with/without graphene layer, the absorption and transmission are illustrated in Figure 3a,b.

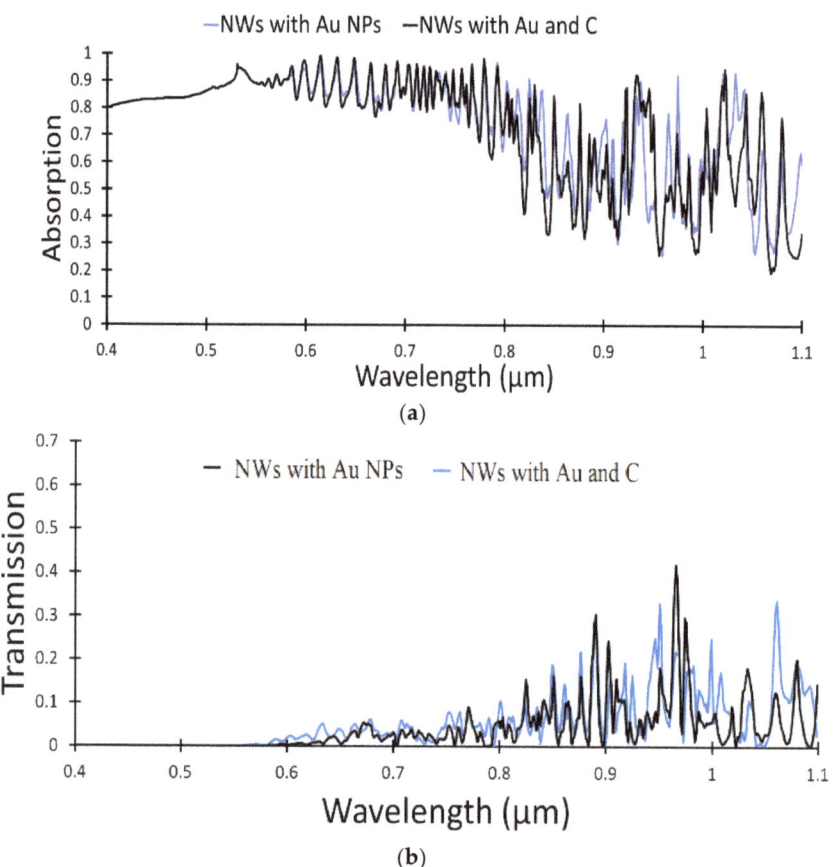

Figure 3. (a) Absorption and (b) transmission of the main two structures: SiNWs with Au NPs distributed on the substrate and SiNWs with graphene layer and Au NPs.

To increase the efficiency of the solar cell, Au NPs are distributed on the top silicon surface directly, without using the graphene layer, and the graphene layer is used to coat the top surface of the SiNWs, as illustrated in Figure 4a,b. The obtained short-circuit current density is 41.7 mA/cm^2 and the overall solar cell efficiency is 19.0%. A 3D structure of the configuration using the graphene layer is shown in Figure 4a, and the side view of the structure is given in Figure 4b. The graphene layer, on the top of NWs, absorbs the incident light and retransmits it into the NWs, which behave as nanoantennas, while the plasmonic Au NPs absorb the light and transfer it to the silicon substrate.

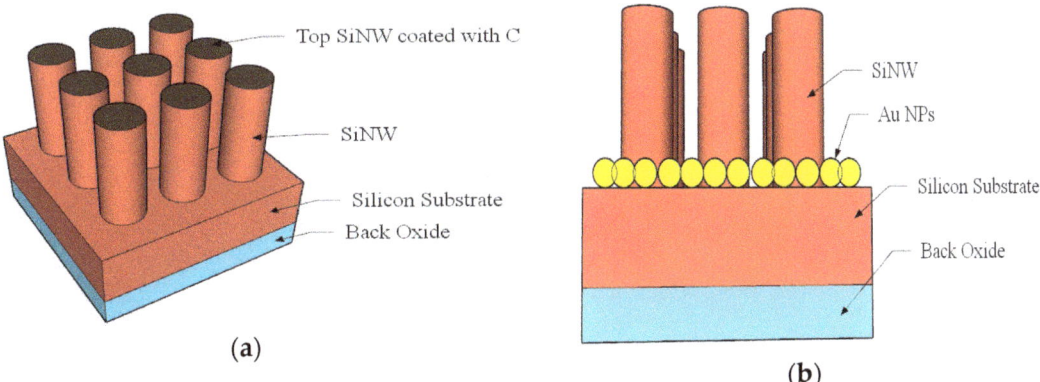

Figure 4. (a) Graphene layer coats the top SiNWs surface; (b) Au NPs distributed on the top of the silicon substrate.

Figure 5 shows the absorption, reflectance, and transmission of the incident light on the proposed SiNW solar cell illustrated in Figure 4.

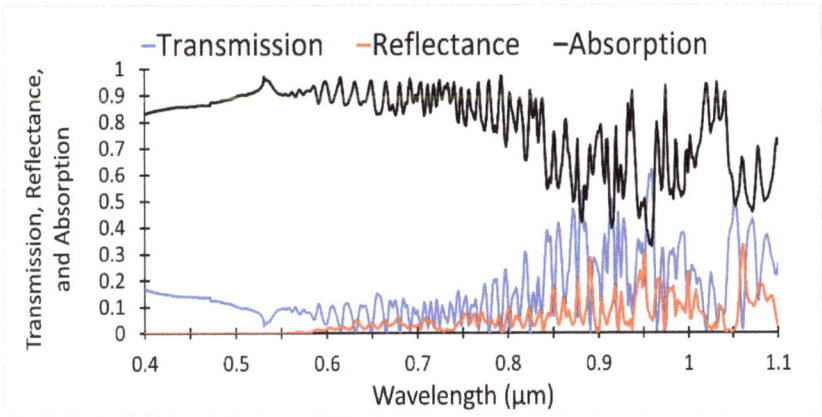

Figure 5. Percentage of absorption, reflectance, and transmission of the SiNWs with a graphene layer on the top of NWs and Au NPs distributed on the silicon substrate.

The reflected and transmitted optical power have been obtained from the simulation analysis by using frequency-domain field and power monitors. Then, the absorption could be calculated using the formula ($A = 1 - T - R$), where T, R, and A are the transmitted, reflected, and absorbed optical power. The absorption is enhanced in the NIR region, which leads to increases in the short-circuit current and power conversion efficiency.

The excess carrier concentration (ECC) has a great effect on the solar cell performance as it determines the open-circuit voltage, FF, and solar cell efficiency. Figure 6 shows the effect of excess carrier concentration on the PCE of the solar cell, which increased with the increase in the excess carrier concentration.

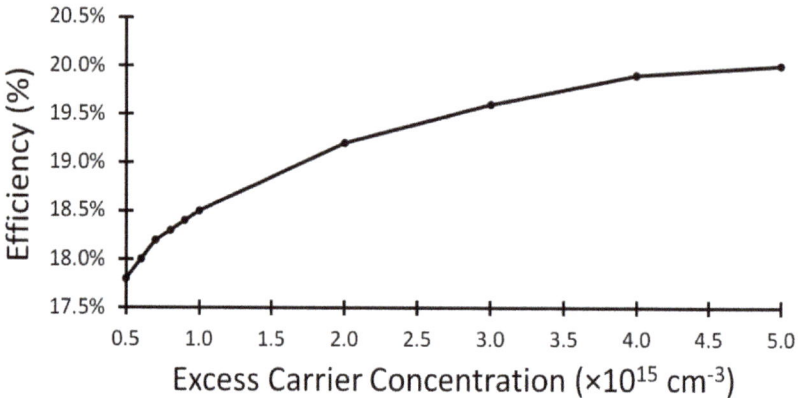

Figure 6. Excess carrier concentration as a function of the PCE for the proposed solar cell structure.

The open-circuit voltage, FF, and PCE are illustrated in Table 3 for different excess carrier concentrations.

Table 3. Open-circuit voltage, FF, and overall efficiency for the proposed structure.

Excess Carrier Concentration ($\times 10^{-15}$ cm^{-3})	V_{oc} (V)	FF	PCE (%)
0.5	0.61	0.72	17.8
0.6	0.61	0.73	18.0
0.7	0.62	0.73	18.2
0.8	0.62	0.73	18.3
0.9	0.62	0.73	18.4
1.0	0.63	0.73	19.0
2.0	0.64	0.73	19.2
3.0	0.66	0.74	19.6
4.0	0.66	0.74	19.9
5.0	0.67	0.74	20.0

As illustrated in Table 3, the efficiency is increased with the increase in the excess carrier concentration and reaches 20% for ECC equal to 5.0×10^{-15} cm^{-3}. However, the ECC value is strongly related to the effective lifetime of the carriers, where the maximum value of the effective lifetime is 3 ms, which is obtained at ECC = 1×10^{-15} cm^{-3}.

The NP material plays a very important role, as illustrated in Equation (6), which consequently changes the J_{sc} and μ values. Figure 7 shows the absorbed energy for different NP materials, namely Au, Ag, Al, and Cu. In the visible region, the absorptions of all NP materials are almost the same, very close to each other, and one cannot distinguish between them, so the main factor in determining the best material is the short-circuit current density and consequently the PCE.

Gold NPs produce a high J_{sc} and PCE, while silver and copper have J_{sc} values equal to 37.6 and 37.4 mA/cm^2, respectively, as clearly given in Table 4. Aluminum NPs give a short-circuit current density of 33.1 mA/cm^2 and overall efficiency of 15.1%. The maximum obtained power conversion efficiency is 19.0% at short-circuit current density J_{sc} = 41.7 mA/cm^2.

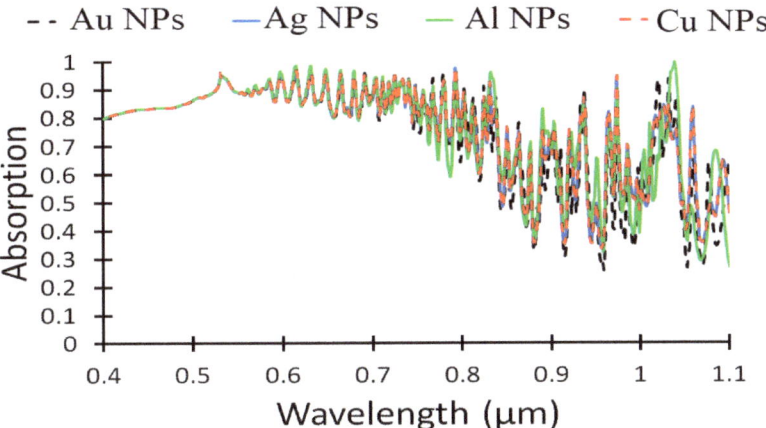

Figure 7. Absorbed energy as a function of wavelength for different NP materials.

Table 4. Short-circuit current density and overall efficiency for different NP materials.

NPs Material	J_{sc} (mA/cm^2)	PCE (%)
Gold, Au	41.7	19.0
Silver, Ag	37.6	17.1
Aluminum, Al	33.1	15.1
Cupper, Cu	37.4	17.0

Plasmonic nanoparticle radius is also considered as one of the main factors that affect solar cell performance. Figure 8 shows the absorbed energy for different Au NP radii, namely 30, 40, 50, and 60 nm. Due to the oscillation of the absorbed power in the NIR region, no distinction between different radii can be considered, so the short-circuit current density is the main factor of choice. The maximum performance, J_{sc} of 41.6 mA/cm^2 and μ of 19.0%, has been obtained at 50 nm Au NP radius, as introduced in Table 5.

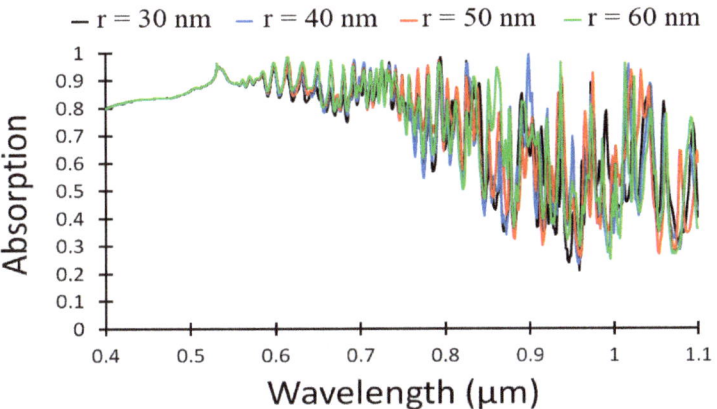

Figure 8. Light absorption (%) as a function of wavelength for different NP radii.

Table 5. Au NP density, short-circuit current density, and overall efficiency for different NP materials.

R (nm)	Au NP Density	J_{sc} (mA/cm^2)	PCE (%)
30	26.0%	34.0	15.5
40	46.0%	34.6	14.4
50	73.0%	41.7	19.0
60	88.4%	33.2	15.1

Figure 9 gives the density of the distributed Au NPs (%) for different radii; the maximum performance of the solar cell is obtained at an NP radius equal to 50 nm, as shown in Table 5.

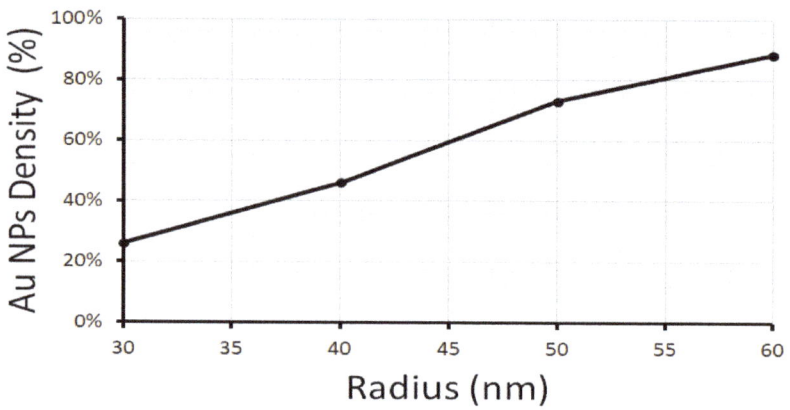

Figure 9. Au NP density for different radii: 30, 40, 50, and 60 nm.

To calculate the Au NP density, the area of all distributed NPs in the unit cell is divided by the available area on the silicon substrate, without NWs.

The period, P, between each NW has a significant effect on the solar cell performance. So, the effect of changing the period from 350 to 600 nm on the short-circuit current density and power conversion efficiency is simulated and summarized in Table 6.

Table 6. Short-circuit current density and overall efficiency for different distances between NWs.

Period (nm)	J_{sc} (mA/cm^2)	PCE (%)
350	33.0	15.0
400	27.8	12.7
450	35.9	16.4
500	40.6	18.5
550	41.7	19.0
600	33.8	15.4

The optimum performance is obtained at period P = 550 nm, which gives J_{sc} = 41.7 mA/cm^2 and PCE = 19.0%. The values of short-circuit current density and PCE decrease as the NW separation distance increases, larger than 500 nm, as the concept of multireflection of the incident light is decreased.

Moreover, the height of the NWs also plays a very important role in solar cell performance. Hence, the effect of NW length on the short-circuit current density and PCE is calculated with a change in the NW length from 3 to 6 µm as given in Table 7. The best performance has been obtained at 5 µm length, as the increase in the length will increase the electron relaxation due to changing the NW bandgap.

Table 7. Short-circuit current density and overall efficiency for different NW lengths.

NW height (μm)	J_{sc} (mA/cm^2)	PCE (%)
3	30.5	13.9
3.5	32.5	14.9
4	33.4	15.2
4.5	37.6	17.1
5	41.7	19.0
5.5	40.2	18.3
6	39.0	17.8

5. Optical Power Distribution

Optical power distribution is monitored at different layers, between the silicon substrate and back oxide, layer 1; at the top of Si substrate, layer 2; in the middle of the SiNWs, layer 3; and at the top of NWs, layer 4. Figure 10 shows the locations of the four layers.

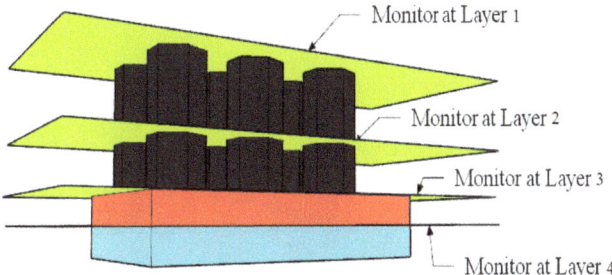

Figure 10. Different positions of electric field monitoring layers.

Figure 11 illustrates the optical power distribution on the four different monitoring layers for all proposed solar cells, traditional SiNWs, SiNWs with C layer, SiNWs with AuNPs, and SINWs with C and Au NPs.

Figure 11a, monitoring layer 1, shows the effect of the graphene layer on the top of NWs; it behaves like a nanoantenna, which retransmits the optical power into the active layer. Au NPs absorb the optical power on the substrate surface, as clearly shown in monitoring 2 with Au NPs; however, monitoring layers 3 and 4 have the same effect.

Finally, the proposed solar cell structure is compared to other structures in the literature, fabricated structures as in [1,5,6,13] and analytical analysis work as in [14], as illustrated in Table 8. The proposed structure does not give the maximum open-circuit voltage, 0.63 V, or *FF*, 0.73; however, it produces a high short-circuit current density, 41.7 mA/cm^2, and consequently a high value of the power conversion efficiency, 19.0%.

Table 8. Comparison between other solar cells and the proposed structure.

Different Structure	V_{oc} (V)	FF	J_{sc} (mA/cm^2)	PCE (%)
[5]	0.56	0.61	17.3	5.31
[6]	0.47	0.59	13.7	3.80
[1]	0.82	0.73	15.2	9.30
[13]	0.51	0.61	33.6	11.61
[14]	0.76	0.86	38.8	-
Proposed structure	0.63	0.73	41.7	19.00

In brief, the main advantage of this study is that the proposed structure enhances the absorption of the SiNWs in the NIR region in addition to the visible region, which leads to improvement in the overall solar cell performance.

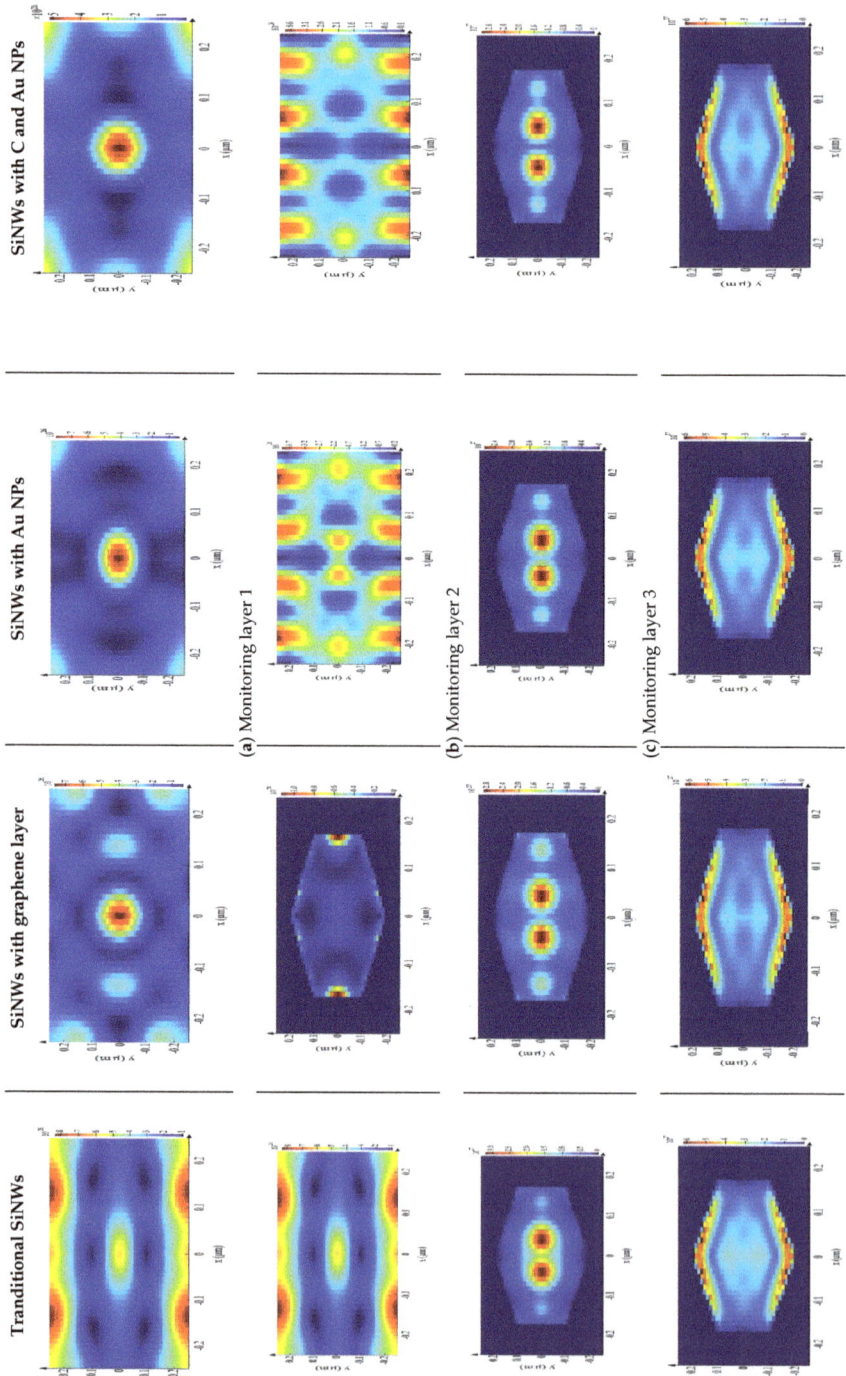

Figure 11. Optical power distribution at different monitoring layers for four different proposed structures.

6. Conclusions

In the proposed solar cell, the top surface of the SiNWs is coated with a graphene layer and the plasmonic Au NPs are distributed on the silicon surface. An FDTD method is used to calculate the optical power absorption and short-circuit current density of a SiNW solar cell. The open-circuit voltage, fill factor, and power conversion efficiency are calculated using numerical analysis. The maximum obtained optical power efficiency is 19.0% for J_{sc} = 41.7 mA/cm^2, V_{oc} = 0.63 V, and FF = 0.73. In addition, the optimum Au NP radius is 50 nm and the excess carrier concentration value is 1×10^{-15} cm^{-3}. The best performance of the introduced solar cell is obtained when Au NPs are used as a plasmonic material, rather than Ag, Al, or Cu materials. Moreover, the optimum NW length is simulated, L = 5 µm, and the distance between two NWs, P = 550 nm, is also calculated. In addition, optical power distribution inside the NWs, on the silicon substrate, and in the interface between substrate and oxide layer shows that the graphene layer and Au NPs behave as nanoantennas by retransmitting the trapped light from the surface into the active layer of the solar cell.

Funding: This research received no external funding.

Institutional Review Board Statement: Not applicable.

Informed Consent Statement: Not applicable.

Data Availability Statement: Not applicable.

Conflicts of Interest: There is no conflict of interest.

References

1. Yu, L.; Cabarrocas, P. Polymorphous nano si and radial junction solar cells. In *Handbook of Photovoltaic Silicon*; Springer: Berlin/Heidelberg, Germany, 2018.
2. Street, R.; Qi, P.; Lujan, R.; Wong, W. Reflectivity of disordered silicon nanowires. *Appl. Phys. Lett.* **2008**, *93*, 163109. [CrossRef]
3. Street, R.; Wong, W.; Paulson, C. Analytic model for diffuse reflectivity of silicon nanowire mats. *Nano Lett.* **2009**, *9*, 3494–3497. [CrossRef] [PubMed]
4. Brönstrup, G.; Jahr, N.; Leiterer, C.; Csáki, A.; Fritzsche, W.; Christiansen, S. Optical properties of individual silicon nanowires for photonic devices. *ACS Nano* **2010**, *4*, 7113–7122. [CrossRef] [PubMed]
5. Garnett, E.; Yang, P. Light trapping in silicon nanowire solar cells. *Nano Lett.* **2010**, *10*, 1082–1087. [CrossRef]
6. Wallace, S.; Jevasuwan, W.; Fukata, N. Silicon nanowires covered with on-site fabricated nanowire-shape graphene for schottky junction solar cells. *Sol. Energy* **2021**, *224*, 666–671. [CrossRef]
7. Li, X.; Zhu, H.; Wang, K.; Cao, A.; Wei, J.; Li, C.; Jia, Y.; Li, Z.; Li, X.; Wu, D. Graphene-on-silicon Schottky junction solar cells. *Adv. Mater.* **2010**, *22*, 2743–2748. [CrossRef]
8. Liu, X.; Zhang, W.; Meng, H.; Yin, G.; Zhang, Q.; Wang, L.; Wu, L. High efficiency schottky junction solar cells by co-doping of graphene with gold nanoparticles and nitric acid. *Appl. Phys. Lett.* **2015**, *106*, 233901. [CrossRef]
9. Elrashidi, A. Electrophotonic improvement of polymer solar cells by using graphene and plasmonic nanoparticles. *Mater. Express* **2017**, *7*, 1–7. [CrossRef]
10. Huang, Y.; Liang, H.; Zhang, Y.; Yin, S.; Cai, C.; Liu, W.; Jia, T. Vertical tip-to-tip interconnection p–n silicon nanowires for plasmonic hot electron-enhanced broadband photodetectors. *ACS Appl. Nano Mater.* **2021**, *4*, 1567–1575. [CrossRef]
11. Jbira, E.; Derouiche, H.; Missaoui, K. Enhancing effect of silver nanoparticles (AgNPs) interfacial thin layer on silicon nanowires (SiNWs)/PEDOT: PSS hybrid solar cell. *Sol. Energy* **2020**, *211*, 1230–1238. [CrossRef]
12. Shin, D.; Kim, J.; Kim, H.J.; Jang, C.; Seo, S.; Lee, H.; Kim, S.; Choi, S. Graphene/porous silicon Schottky-junction solar cells. *J. Alloys Compd.* **2017**, *715*, 291–296. [CrossRef]
13. Yoon, J.; Baca, A.; Park, S.; Elvikis, P.; Geddes, J.; Li, L.; Kim, R.; Xiao, J.; Wang, S.; Kim, T.; et al. Ultrathin silicon solar microcells for semitransparent, mechanically flexible and microconcentrator module designs. *Nat. Mater.* **2008**, *7*, 907–915. [CrossRef] [PubMed]
14. Augusto, A.; Herasimenka, S.; King, R.; Bowden, S.; Honsberg, C. Analysis of the recombination mechanisms of a silicon solar cell with low bandgap-voltage offset. *J. Appl. Phys.* **2017**, *121*, 205704. [CrossRef]
15. Zhou, F.; Qin, F.; Yi, Z.; Yao, W.; Liu, Z.; Wu, X.; Wu, P. Ultra-wideband and wide-angle perfect solar energy absorber based on Ti nanorings surface plasmon resonance. *Phys. Chem. Chem. Phys.* **2021**, *23*, 17041–17048. [CrossRef]
16. Deng, Y.; Cao, G.; Wu, Y.; Zhou, X.; Liao, W. Theoretical Description of Dynamic Transmission Characteristics in MDM Waveguide Aperture-Side-Coupled with Ring Cavity. *Plasmonics* **2015**, *10*, 1537–1543. [CrossRef]

17. Deng, Y.; Cao, G.; Yang, H.; Zhou, X.; Wu, Y. Dynamic Control of Double Plasmon-Induced Transparencies in Aperture-Coupled Waveguide-Cavity System. *Plasmonics* **2018**, *13*, 345–352. [CrossRef]
18. Zheng, Z.; Zheng, Y.; Luo, Y.; Yi, Z.; Zhang, J.; Liu, Z.; Yang, W.; Yu, Y.; Wu, X.; Wu, P. A switchable terahertz device combining ultra-wideband absorption and ultra-wideband complete reflection. *Phys. Chem. Chem. Phys.* **2022**, *24*, 2527–2533. [CrossRef]
19. Zhao, F.; Lin, J.; Lei, Z.; Yi, Z.; Qin, F.; Zhang, J.; Liu, L.; Wu, X.; Yang, W.; Wu, P. Realization of 18.97% theoretical efficiency of 0.9 µm thick c-Si/ZnO heterojunction ultrathin-film solar cells via surface plasmon resonance enhancement. *Phys. Chem. Chem. Phys.* **2022**, *24*, 4871–4880. [CrossRef]
20. Yang, X.; Wang, D. Photocatalysis: From Fundamental Principles to Materials and Applications. *ACS Appl. Energy Mater.* **2018**, *1*, 6657–6693. [CrossRef]
21. Nan, T.; Zeng, H.; Liang, W.; Liu, S.; Wang, Z.; Huang, W.; Yang, W.; Chen, C.; Lin, Y. Growth Behavior and Photoluminescence Properties of ZnO Nanowires on Gold nano-particle Coated Si Surfaces. *J. Cryst. Growth* **2012**, *340*, 83–86. [CrossRef]
22. Kang, Z.; Gu, S.; Yan, Q.; Bai, M.; Liu, H.; Liu, S.; Zhang, H.; Zhang, Z.; Zhang, J.; Zhang, Y. Enhanced Photoelectrochemical Property of ZnO Nanorods Array Synthesized on Reduced Graphene Oxide for Self-powered Biosensing Application. *Biosens. Bioelectron.* **2015**, *64*, 499–504. [CrossRef] [PubMed]
23. Kang, Z.; Yan, Q.; Wang, F.; Bai, M.; Liu, C.; Zhang, Z.; Lin, P.; Zhang, H.; Yuan, G.; Zhang, J.; et al. Electronic Structure Engineering of Cu_2O Film/ZnO Nanorods Array All-oxide p-n Heterostructure for Enhanced Photoelectrochemical Property and Self-powered Biosensing Application. *Sci. Rep.* **2015**, *5*, 7882. [CrossRef] [PubMed]
24. Sinton, R.; Cuevas, A. Contactless determination of current–voltage characteristics and minority-carrier lifetimes in semiconductors from quasi-steady-state photoconductance data. *Appl. Phys. Lett.* **1996**, *69*, 2510. [CrossRef]
25. Aspnes, D.; Studna, A. Dielectric functions and optical parameters of Si, Ge, GaP, GaAs, GaSb, InP, InAs, and InSb from 1.5 to 6.0 eV. *Phys. Rev.* **1983**, *27*, 985. [CrossRef]
26. Phillip, H.; Taft, E. Kramers-Kronig Analysis of Reflectance Data for Diamond. *Phys. Rev.* **1964**, *136*, A1445. [CrossRef]
27. Elrashidi, A.; Tharwat, M. Broadband absorber using ultra-thin plasmonic metamaterials nanostructure in the visible and near-infrared regions. *Opt. Quantum Electron.* **2021**, *53*, 426. [CrossRef]

MDPI
St. Alban-Anlage 66
4052 Basel
Switzerland
Tel. +41 61 683 77 34
Fax +41 61 302 89 18
www.mdpi.com

Applied Sciences Editorial Office
E-mail: applsci@mdpi.com
www.mdpi.com/journal/applsci

www.ingramcontent.com/pod-product-compliance
Lightning Source LLC
LaVergne TN
LVHW070543100526
838202LV00012B/367